Memoirs of the American Mathematical Society

Number 202

Irwin E. Schochetman

Kernels and integral operators for continuous sums of Banach spaces

Published by the

AMERICAN MATHEMATICAL SOCIETY

Providence, Rhode Island

VOLUME 14 · NUMBER 202 (end of volume) · MAY 1978

ABSTRACT

The purpose of this Memoir is twofold: 1. To extend the notions of kernel and integral operator to the context of continuous sums of Banach spaces. 2. To extend to this setting, certain well-known conditions for an integral operator to be compact, Hilbert-Schmidt and trace-class.

AMS (MOS) Subject Classifications (1970)

Primary: 47B05, 48B10, 47G05

Secondary: 22D12, 22D20, 22D30, 28A45, 46E30

Key Words and Phrases

Vector field, continuity structure, equi-continuity, continuous sum of Banach spaces, dual continuity structure kernel field, integral operator, induced Banach representation, compactness.

Library of Congress Cataloging in Publication Data **CIP**

Schochetman, Irwin E., 1939-
Kernels and integral operators for continuous sums of Banach spaces.

(Memoirs of the American Mathematical Society ; no. 202)
Bibliography: p.
1. Banach spaces. 2. Integral operators. 3. Kernel functions. I. Title. II. Title: Continuous sums of Banach spaces. III. Series: American Mathematical Society. Memoirs ; no. 202.
QA3.A57 no. 202 [QA323] 510'.8s [515'.73] 78-4580
ISBN 0-8218-2202-0

CONTENTS

PREFACE

These notes were written for the purpose of extending the concepts of integral operator and kernel (and their properties) to the context of continuous sums of Banach spaces. Such spaces were introduced by R. Godement in 1951. The Hilbert space version of this extension was carried out by M. Duflo in 1972. Thus, my objective here is to complete this extension to the general Banach space setting.

The reader is required to know only the basics of: (1) point-set topology of locally compact Hausdorff spaces, (2) integration theory of vector-valued functions on such spaces, (3) operator theory of Banach spaces. Just about everything else is developed in the notes.

The major part of this work was done while I was on sabbatical leave at the University of Pennsylvania in the winter and spring of 1975. I am indebted to Penn's Mathematics Department for extending this privilege to me, as well as for its generosity and hospitality. I am especially grateful in this regard to J. M. G. Fell and R. V. Kadison.

While working on these notes, I benefited (at Penn and at Oakland) from conversations with several colleagues--notably: R. Boyer, R. A. Fontenot, R. Martin, C. Radin, and R. Sharpley. I am particularly indebted to Professor Fontenot for the proof of a very important "one-to-one-ness" result on the correspondence between integral operators and kernels.

These notes were typed by Ava Kerr; a special thank you is due her for doing such a nice job. I also wish to express my appreciation to Lewis Pino (Director of Oakland University's Office of Research Services) for making her services available to me.

Rochester, Michigan
July, 1976

KERNELS AND INTEGRAL OPERATORS FOR CONTINUOUS SUMS OF BANACH SPACES*

0. INTRODUCTION, BACKGROUND, AND PRELIMINARIES

Suppose X is a locally compact Hausdorff space, μ is a positive Radon measure on X, and E is a Banach space. For functions from X into E, there exists a canonical meaning for the notion of continuity and, consequently, for μ-measurability. These in turn give rise to the familiar Lebesque spaces $L^p(X,\mu,E)$. As is well-known, an integral operator on $L^p(X,\mu,E)$ is a bounded linear operator T for which there exists a kernel, i.e. a mapping from $X\times X$ into the bounded operators on E which in some sense: (1) is measurable relative to μ and (2) satisfies the formula

$$(Tf)(x) = \int_X K(x,y)f(y)d\mu(y) ,$$

for f belonging to at lease a dense subspace of $L^p(X,\mu,E)$. (Of course, there are questions concerning integrability and measurability which have to be resolved.) For the kernel K, there exist familiar notions of continuity and measurability--namely, weak, strong, and uniform. Moreover, there exist results (partial or complete) which determine under what conditions on K the operator T is compact, Hilbert-Schmidt, or trace-class.

Now suppose the space E is replaced by a field of Banach spaces over X, i.e. a collection of Banach spaces $\{E_x\}$ indexed by X. The "mappings" we want to consider here are the vector fields, i.e. the elements of the product space $\prod_X E_x$. In general, there is no canonical meaning for continuity of these objects. In order to study this concept for vector fields, we assume we are also given a "continuity structure" Λ in $\prod_X E_x$ with respect to which

*Received by the editors January 4, 1977.

continuity (and hence measurability) can be defined (Godement [1]). Relative to such a structure Λ , R. Godement extended the concept of Lebesque space to the vector field setting by constructing the spaces $L^p(\Lambda,\mu)$ of p-integrable "vector fields." It is these spaces which are called "continuous sums of Banach spaces."

It is very natural (as well as very interesting) to want to also extend the notions and properties of integral operator and kernel to the vector field setting. In a nutshell, this is our objective in these notes. We were very strongly motivated to do this by the recent work, Duflo [1], of M. Duflo. Here, he substantially accomplished the Hilbert space version of our objective in his attempt to characterize the Hilbert-Schmidt and trace-class properties for integral operators having weakly measurable or weakly continuous kernels. This added further incentive to carrying out the above extension to the general Banach space setting. Accordingly, we have concentrated on (1) developing the notions of weak, strong, and uniform continuity and measurability for Banach-type kernels and (2) studying the compactness of integral operators on the spaces $L^p(\Lambda,\mu)$.

In addition to the above reasons, the author had some more practical reasons for accomplishing the above objective--as did Duflo. In his study of induced unitary representations U of locally compact groups, Duflo was able to write each operator of the integrated form of U as an integral operator on a Hilbert space $L^2(\Lambda,\mu)$ having weakly continuous kernel. In this way, he studied the Hilbert-Schmidt and trace-class properties of the representation U. Analogously, in studying induced Banach representations (Fontenot and Schochetman [1]), the author has been able to show that Duflo's construction can be extended to yield integral operators on a Banach space $L^p(\Lambda,\mu)$ (Schochetman [1]). Furthermore, it has been known intuitively for some time (Schochetman [2] Prop. 4.2) that the associated kernels should be "uniformly continuous." Hence, one of our specific objectives was to formalize the meaning of this property in the vector-field setting and verify that it was true in this application. Another specific objective was to extend a useful kernel condition on compactness of integral operators so as to be able to apply it to the operators of the integrated form of an induced Banach representation.

The remaining chapters and sections are organized as follows. Chapter I is devoted mainly to the study of continuity structures and continuity (section 1) and measurability (section 3). In conjunction with continuity, we also study equicontinuity (section 2) for families of vector fields. In section 4, we establish some needed facts about continuity structures on product spaces.

Chapter II is concerned with spaces of vector fields. In section 5, we study spaces of continuous vector fields having the usual additional properties, e.g. compact support, etc. Section 6 contains the construction of the Lebesque spaces. Of particular significance is the characterization of when $L^p(\Lambda,\mu)$ is a Hilbert space. Section 7 repeats the previous construction to obtain the mixed-norm Lebesque spaces over a product space. In section 8, duality between continuity structures and Lebesque spaces is studied.

The notion of kernel is introduced and studied in Chapter III. After establishing two useful norm formulas for a kernel in section 9, the definitions and properties of weak and strong continuity and measurability are developed in sections 10 and 11 respectively. Section 12 contains our treatment of uniform continuity and norm continuity for kernels--these turn out to be slightly different in the vector field setting. The last two sections in this chapter study certain types of kernels--namely, those which are of finite-rank (section 13) or Hilbert-Schmidt (section 14).

Chapter IV, the chapter on integral operators, brings together all the previous results. In section 15 we develop the notion of integral operator relative to a kernel and study the correspondence between these objects. Sections 16, 17, and 18 are devoted to finding kernel conditions which imply the integral operator is compact, Hilbert-Schmidt or trace-class respectively.

In Chapter V we apply the results of the first four chapters to the integrated form of an induced Banach representation. Section 19 contains the construction of such a representation. In section 20 we describe how the operators of the integrated form may be viewed as integral operators with norm-continuous kernels. In section 21, we translate the main results of sections 16, 17, and 18 to the context of an induced representation.

At the ends of some of the sections, we have included certain questions which we considered but did not answer (for one reason or another).

It will be convenient throughout these notes to adopt the following notational conventions. Let X be a topological space, D a subset of X, and E a Banach space. If $f:X\to E$, then $\mathrm{supp}(f)$ will denote the support of f in X and we define:

$$\|f\|_D = \sup_{x\in D}\|f(x)\| .$$

The characteristic function of D will be denoted by γ_D. If x is a point in D, then $C(x,D,E)$ will denote the space of functions from D into E which are relatively continuous at x. Moreover,

$$C(D,E) = \bigcap_{x\in D} C(x,D,E) .$$

The subspaces of $C(X,E)$ consisting of functions which are bounded, vanish at infinity, or have compact support will be denoted by $C_b(X,E)$, $C_0(X,E)$, and $C_c(X,E)$ respectively. If E equals the complex numbers $\mathbb{C}$, we will delete it. Recall that the spaces $C(x_0,D)$, $C(D)$, and $C_b(D)$ are rings relative to the pointwise operations If $\{A_x\}$ is a family of sets indexed by X then ΠA_x will mean that the Cartesian product is taken over all of X. If the product is taken over the subset D , we will write (as usual) $\prod_D A_x$. For the rest of these notes, X will denote a fixed locally compact Hausdorff space and $\mathbb{R}$ will be the real numbers.

A measure μ on X will automatically be a positive Radon measure and supp(μ) its support. In writing an integral we will usually delete references to the measure (if there is no possibility of confusion) and the domain (only if it's the whole space). Thus, roughly,

$$\int f(x)dx = \int_X f(x)d\mu(x).$$

The space of μ-measurable functions from X into E will be denoted by $M(X,\mu,E)$. Again, E will be deleted if it is equal to $\mathbb{C}$. As above, the space $M(X,\mu)$ is a ring relative to the pointwise operations.

If E^* denotes the topological dual of E , then $\langle v,v^*\rangle$ will denote the value of the functional v^* of E^* at the vector v of E. If F is another Banach space, then Hom(F,E) will denote the Banach space of bounded linear operators from F into E and $\mathrm{Hom}_\infty(F,E)$ the closed subspace of compact operators. If, in particular, E and F are Hilbert spaces, then $\mathrm{Hom}_2(F,E)$ will denote the Hilbert space of Hilbert-Schmidt operators from F into E and $\|\cdot\|_s$ the norm in this space. If E = F , then we will write Hom(E) for Hom(E,E) , etc. (whether E is a Hilbert space or not). The notation tr(A) will denote the trace of a trace-class operator A on the Hilbert space E.

Throughout these notes, we assume $\{E_x\}$ is a fixed given field of Banach spaces over the locally compact Hausdorff space X.

I. CONTINUOUS AND MEASURABLE VECTOR FIELDS

Suppose for the moment that the field $\{E_x\}$ of Banach spaces over X is constant, i.e. $E_x = E$, for each x in X. Thus, the vector fields are just the functions from X into E. In this case, since X and E are topological spaces there is a very familiar notion of continuity which is derived from the existence of an "inherently continuous" family of functions, namely, the constant functions. Basically, a function $f: X \to E$ is continuous at a point x_o in X if its values are arbitrarily close to the value of a constant function in a neighborhood of x_o. Once continuity has been defined, the notions of equicontinuity and measurability are then derived from it in a familiar fashion. The same is true for the product structures.

Now suppose the field of Banach spaces $\{E_x\}$ is again arbitrary. Our objective in this chapter is to accomplish the above in this context, i.e., for vector fields. In order to do this, we must start with an "inherently continuous" collection of vector fields in ΠE_x upon which to build the concept of continuity. Such a collection was originally called "a fundamental family of continuous vector fields" in R. Godement[1] and later a "continuity structure" in J.M.G. Fell [1]. We will use the latter terminology since it seems to reflect the essence of the object more accurately. Unfortunately, as we shall see (1.25), continuity structures don't exist in general. Hence, their existence will have to be assumed. A general reference for this chapter is Godement [1]. In addition, particular references for section 1 are Dixmier [1,§10] and Fell [1].

1. CONTINUITY STRUCTURES AND CONTINUITY

In this section, we introduce the notion of continuity for vector fields (relative to a given continuity structure) and study many of its basic properties.

1.1 DEFINITION. A <u>continuity structure</u> in ΠE_x is a subset Λ satisfying:

(i) Λ is a vector subspace of ΠE_x.

(ii) For each h in Λ, the scalar function $||h||$ is continuous on X.

(iii) For each x in X, the subspace of E_x given by $\{h(x):h\epsilon\Lambda\}$ is dense.

On occasion, we will be interested in the particular case where there exists a countable subset of Λ satisfying (iii) above. When such exists, we will say that Λ is <u>countably dense</u>.

1.2 EXAMPLE. If $E_x=E$, for all x in X, then the family Λ_0 of constant functions from X into E is a continuity structure in ΠE_x. More generally, suppose $\omega:X\to(0,\infty)$ is a continuous function and E_x is defined to be the Banach space equivalent to E with norm equal to that of E multiplied by $\omega(x)$. Then Λ_0 is still a continuity structure in ΠE_x.

1.3 EXAMPLE. If X is discrete and $\{E_x\}$ is arbitrary, then $\Lambda=\Pi E_x$ is a continuity structure. Actually, in this case, this is essentially the only one (1.22).

Throughout the rest of this manuscript, we will assume Λ is a given continuity structure in ΠE_x. Next we discuss continuity relative to Λ for vector fields.

1.4 DEFINITION. Let D be a subset of X, x_0 an element of D and f a vector field defined on D, i.e. $f\epsilon\Pi_D E_x$. Then f is <u>Λ-continuous</u> at x_0 if given $\varepsilon>0$, there exists a neighborhood N of x_0 in X and an element of h of Λ such that $||f(x)-h(x)||<\varepsilon$, for all x in $N\cap D$. Denote the set of such vector fields by $C(x_0,D,\Lambda)$. The field f is Λ-continuous on D if it is Λ-continuous at each point of D. Denote such by $C(D,\Lambda)$; if D=X, we write $C(\Lambda)$ for $C(X,\Lambda)$. Clearly, $\Lambda\subseteq C(\Lambda)$ in general. For the sake of convenience, when there is no possibility of confusion, we will delete the reference to Λ in the terminology, but not in the notation. For example, f is continuous will mean f is in $C(\Lambda)$.

1.5 EXAMPLES. (1) If Λ_0 is as in 1.2, then $C(\Lambda_0)=C(X,E)$. (2) If Λ is as in 1.3, then $C(\Lambda)=\Lambda=\Pi E_x$; in fact, if Λ is any continuity structure in ΠE_x, X discrete, then $C(\Lambda)=\Pi E_x$ by (iii) of 1.1.

Let D be a subset of X and x_0 an element of D. The following properties of a continuity structure are easily verified:

1.6 LEMMA. Continuity is preserved under restriction. Specifically, let D' be a subset of X such that $D \subseteq D'$. If $f \in C(x_0,D,\Lambda)$, then $f|D' \in C(x_0,D',\Lambda)$. Thus, $f \in C(D,\Lambda)$ implies $f|D' \in C(D',\Lambda)$.

1.7 LEMMA. The mapping $f \to \|f\|$ preserves continuity. Specifically, if $f \in C(x_0,D,\Lambda)$, then $\|f\| \in C(x_0,D)$. Thus, $f \in C(D,\Lambda)$ implies $\|f\| \in C(D)$.

1.8 PROPOSITION. Continuity is preserved under multiplication by continuous scalar functions. Specifically, if $f \in C(x_0,D,\Lambda)$ and $\phi \in C(x_0,D)$, then $\phi f \in C(x_0,D,\Lambda)$.

Proof. Let $\varepsilon>0$. By hypothesis, it follows from 1.7 that there exists a neighborhood N' of x_0 in X such that

$$\big|\, \|f(x)\| - \|f(x_0)\| \,\big| < 1,$$

i.e.

$$\|f(x)\| \leq 1 + \|f(x_0)\|, \qquad x \in N' \cap D.$$

Moreover, there exists a neighborhood $N'' \subseteq N'$ of x_0 in X and h in Λ such that

$$\|f(x)-h(x)\| < \frac{\varepsilon}{2(|\phi(x_0)|+1)}, \qquad x \in N'' \cap D.$$

Since ϕ is continuous at x_0, there exists a neighborhood $N \subseteq N''$ of x_0 in X such that

$$|\phi(x)-\phi(x_0)| \leq \frac{\varepsilon}{2(\|f(x_0)\|+1)}, \qquad x \in N \cap D.$$

Thus:

$$\begin{aligned}\|\phi(x)f(x)-\phi(x_0)h(x)\| &\leq \|\phi(x)f(x)-\phi(x_0)f(x)\| + \|\phi(x_0)f(x)-\phi(x_0)h(x)\| \\ &\leq |\phi(x)-\phi(x_0)|\,\|f(x)\| + (|\phi(x_0)|+1)\|f(x)-h(x)\| \\ &\leq \varepsilon, \qquad\qquad x\in N\cap D.\end{aligned}$$

Since $\phi(x_0)h$ is in Λ, we have that $\phi f\in C(x_0,D,\Lambda)$.

1.9 COROLLARY. (i) The space $C(x_0,D,\Lambda)$ is a complex linear space and a $C(x_0,D)$- module. (ii) Similarly for $C(D,\Lambda)$ and $C(D)$.

1.10 LEMMA. Let f be a vector field defined on D. Then $f\in C(x_0,D,\Lambda)$ if and only if for each h in Λ, the scalar function on D given by $x\to\|f(x)-h(x)\|$ is continuous at x_0.

Now suppose H is a subset of Λ and Φ a subset of $C(X)$. Let $\Phi\cdot H$ denote the linear span of $\{\phi h : \phi\in\Phi,\ h\in H\}$ in $C(\Lambda)$ (1.9). The next result characterizes the elements of $C(\Lambda)$ in terms of $C(X)$ and Λ.

1.11 THEOREM. Let f be a vector field. Then $f\in C(\Lambda)$ if and only if on each compact subset of X, f is a uniform limit of elements of $C(X)\cdot\Lambda$. Furthermore, if Λ is countably dense with H as in 1.1, then $C(X)\cdot\Lambda$ may be replaced by $C(X)\cdot H$ in the previous statement.

Proof. This is essentially Proposition 6 of Godement [1,p.81].

1.12 COROLLARY. If the net $\{f_n\}$ in $C(\Lambda)$ converges to the vector field f uniformly on X, i.e. $\|f_n-f\|_X\to 0$, then $f\in C(\Lambda)$.

If D happens to be a locally compact subspace of X, then we define $\Lambda|D=\{h|D : h\in\Lambda\}$. It is easy to see that $\Lambda|D$ is a continuity structure in $\prod_D E_x$ satisfying $C(\Lambda|D)=C(D,\Lambda)$. Also, if Λ is countably dense by $H\subseteq\Lambda$, then $\Lambda|D$ is countably dense by $H|D$.

In addition to countable density, there are other separability notions we will require.

1.13 DEFINITION. Let F be a subset of $C(\Lambda)$. Then F is <u>total</u> if for each x in X, the set $\{f(x) : f\varepsilon F\}$ is total in E_x. The continuity structure Λ is <u>separable</u> if there exists a countable total subset of $C(\Lambda)$. (By taking linear combinations with complex rational coefficients, we may assume $\{f(x) : f\varepsilon F\}$ is dense in E_x.) In Section 15, we will need a local version of separability. The continuity structure Λ is <u>locally separable</u> if for each compact subset Q of X, the continuity structure $\Lambda|Q$ is separable.

1.14 REMARKS. (1) Λ is total. (2) If Λ is countably dense, it is separable. (3) If Λ is separable, it is locally separable. (4) The converses of (2) and (3) are false. To see this in the case of (2), let X be an uncountable discrete space and E_x the complex numbers, for each x in X. Let

$$\Lambda=\{f:X\to E : f(x)=0, \text{ for all but finitely many } x\}.$$

Then Λ is a continuity structure in ΠE_x; in fact, $C(\Lambda)=\Pi E_x$ as we shall see (1.22). Moreover, $C(\Lambda)=C(X)$, so that Λ is separable. However, it is clear that Λ is not countably dense.

Although a continuity structure Λ is dense fiberwise, each fiber is completely covered by $C(\Lambda)$ in this way. Specifically:

1.15 LEMMA (Urysohn). Let Q be a compact subset of X and f an element of $C(\Lambda|Q)$. Then there exists g in $C(\Lambda)$ such that $g|Q=f$.

1.16 COROLLARY. If $Q=\{x_1,\ldots,x_n\}$ and v_i is an element of E_i, $i=1,\ldots,n$, then there exists f in $C(\Lambda)$ such that $f(x_i)=v_i, i=1,\ldots,n$.

Next we investigate equivalence between continuity structures. Intuitively speaking, Λ is significant only to the extent that it gives rise to $C(\Lambda)$.

1.17 PROPOSITION. The space $C(\Lambda)$ is a continuity structure in ΠE_x and $C(C(\Lambda))=C(\Lambda)$.

Proof. Properties (i) and (ii) of 1.1 follow from 1.9 and 1.7 respectively. Property (iii) is clear since $\Lambda \subseteq C(\Lambda)$; in fact, $\{f(x): f \in C(\Lambda)\} = E_x, x \in X$, by 1.16. Thus, we need only show that $C(C(\Lambda)) \subseteq C(\Lambda)$. However, this follows by application of the triangle inequality.

1.18 REMARKS. This proposition shows: (1) There exist different continuity structures for a given field of Banach spaces. (2) Different structures may yield the same continuous vector fields. This suggests comparing different Λ's according to their corresponding $C(\Lambda)$'s.

For the following, let Λ and Λ' be two continuity structures in ΠE_x.

1.19 LEMMA. Let D be a subset of X and x_0 an element of D. Suppose $\Lambda \subseteq C(\Lambda')$. If the vector field f defined on D is Λ-continuous at x_0, then it is Λ'-continuous at x_0, i.e. $C(x_0, D, \Lambda) \subseteq C(x_0, D, \Lambda')$. Consequently, if $\Lambda \subseteq C(\Lambda')$, then $C(D, \Lambda)$ is contained in $C(D, \Lambda')$, for all subsets D of X.

1.20 COROLLARY. The following are equivalent: (i) Λ is contained in $C(\Lambda')$. (ii) $C(D,\Lambda)$ is contained in $C(D,\Lambda')$ for all subsets D of X.

1.21 DEFINITION. We define $\Lambda \leq \Lambda'$ by $C(\Lambda) \subseteq C(\Lambda')$ and $\Lambda \sim \Lambda'$ by $C(\Lambda) = C(\Lambda')$. Clearly, $\leq$ is a partial ordering and $\sim$ is an equivalence relation, the latter being the same as "strict equivalence" defined by Fell in Fell [1, p.236].

1.22 EXAMPLES. (1) In general, $C(\Lambda) \sim \Lambda$ (1.17) but $C(\Lambda) \neq \Lambda$, i.e. the relation $\sim$ is weaker than equality. Note also that if Λ is separable, then it is equivalent to a countably dense continuity structure, namely $C(\Lambda)$.

(2) If $E_x = E$, for all x in X, with Λ_0 as in 1.2, and Λ is a continuity structure in ΠE_x such that $\Lambda \leq \Lambda_0$, i.e. $\Lambda \subseteq C(X,E)$, then $\Lambda \sim \Lambda_0$. This can be seen as follows. Clearly, $C(\Lambda) \subseteq C(X,E)$ by 1.20. On the other hand, let f be an element of $C(X,E)$. If $x_0 \in X$ and $\varepsilon > 0$, then there exists a neighborhood N' of x_0 in X such that $\|f(x) - f(x_0)\| < \varepsilon/3$, $x \in N'$. But $\{h(x_0): h \in \Lambda\}$ is dense in $E_{x_0} = E$. Hence, there exists h in Λ such that $\|f(x_0) - h(x_0)\| < \varepsilon/3$. Since $\Lambda \subseteq C(X,E)$, there exists a

neighborhood $N \subseteq N'$ of x_0 in X such that $\|h(x)-h(x_0)\| < \varepsilon/3$, $x \in N$. Thus, $\|f(x)-h(x)\| < \varepsilon$, $x \in N$, by the triangle inequality, so that $f \in C(\Lambda)$. (3) if X is discrete, then there exists only one $\sim$-equivalence class of continuity structures in ΠE_x (1.5). (4) The previous situation is not the case in general. For example, let $X = \mathbb{R}$ and $E_x = \mathbb{C}$, $x \in \mathbb{R}$. Define

$$f(x) = \begin{cases} 1, & x \geq 0, \\ -1, & x < 0, \end{cases}$$

and $\Lambda = \{cf : c \in \mathbb{C}\}$. Then Λ is a continuity structure in ΠE_x and is inequivalent to Λ_0; in fact, $C(\Lambda) = \{gf : g \in C(\mathbb{R})\}$.

Combining 1.19, 1.20 and 1.21, we obtain:

1.23 PROPOSITION. Let Λ and Λ' be continuity structures in ΠE_x. The following are equivalent:

(i) $\Lambda \sim \Lambda'$.

(ii) $\Lambda \subseteq C(\Lambda')$ and $\Lambda' \subseteq C(\Lambda)$.

(iii) For each subset D of X and x_0 in D, a vector field f on D is Λ-continuous at x_0 if and only if it is Λ'-continuous at x_0.

(iv) For each subset D of X, $C(D,\Lambda) = C(D,\Lambda')$.

The following is a very useful characterization of equivalence.

1.24 PROPOSITION. The continuity structures Λ and Λ' are equivalent if and only if there exists a subset F of $C(\Lambda) \cap C(\Lambda')$ such that $\{f(x) : f \in F\}$ is dense in E_x for each x in X (Fell [1, p.236]).

Finally in this section, we consider the question of existence of a continuity structure for a given field of Banach spaces.

1.25 LEMMA. Suppose each E_x is a Hilbert space, $x \in X$. If there exists a continuity structure Λ in ΠE_x, then for each integer n, the set

$$\{x \in X : \text{dimension of } E_x \leq n\}$$

is closed in X. Conversely, if each E_x is also separable, then this condition is sufficient for the existence of a continuity structure in ΠE_x.

Proof. See Godement [1, p.83].

Therefore, contrary to the situation in the classical (1.2) and discrete (1.3) cases, there need not exist a continuity structure (canonical or otherwise) in ΠE_x in general.

2. EQUICONTINUITY

Having extended the notion of continuity to the context of vector fields, it is natural to do likewise for the notion of equicontinuity. Moreover, our treatment of uniform continuity for kernels in section 12 requires that this be done.

For the following, let D be a subset of X and x_0 an element of D.

2.1 DEFINITION. Let F be a family of vector fields defined on D, i.e. $F \subseteq \prod_D E_x$. Then F is <u>Λ-equicontinuous</u> at x_0 if given $\varepsilon > 0$, there exists a neighborhood N of x_0 in X satisfying the following:
For each f in F, there exists h_f in Λ such that $\|f(x)-h_f(x)\| < \varepsilon$, $x \in N \cap D$.
For convenience, denote the set of such families by $E(x_0,D,\Lambda)$. The family F is Λ-equicontinuous on D if it is Λ-equicontinuous at each point of D. Denote such by $E(D,\Lambda)$; if D=X, delete it. Of course, "$C(\Lambda) \subseteq E(\Lambda)$." If there is no danger of confusion we will also delete reference to Λ in the terminology, but not in the notation.

The next four lemmas are easily verified.

2.2 LEMMA. If F is equicontinuous at x_0, then each f in F is continuous at x_0. Conversely, if the vector field f defined on D is continuous at x_0, then $\{f\}$ is equicontinuous at x_0. In particular, $\{f\} \in E(\Lambda)$, for f in $C(\Lambda)$.

2.3 LEMMA. If H is a subset of Λ, the $H|D$ is equicontinuous on D, i.e. $H|D \in E(D,\Lambda)$. In particular, $H \in E(\Lambda)$.

2.4 LEMMA. If D is a locally compact subspace of X, then $E(\Lambda|D) = E(D,\Lambda)$.

2.5 LEMMA. If F is an element of $E(\Lambda)$, then G belongs to $E(\Lambda)$ for all subsets G of F.

Next we compare different continuity structures in terms of equicontinuity.

We will obtain an equivalence relation which is stronger than the relation $\sim$. Suppose Λ and Λ' are two continuity structures in ΠE_x.

2.6 LEMMA. If $E(\Lambda)$ is contained in $E(\Lambda')$, then $C(\Lambda)$ is contained in $C(\Lambda')$. The converse is false in general.

Proof. If $f\varepsilon C(\Lambda)$, then $\{f\}\varepsilon E(\Lambda)$ by 2.2, so that $\{f\}\varepsilon E(\Lambda')$, i.e. $f\varepsilon C(\Lambda')$. The second part follows from the next familiar example.

2.7 EXAMPLE. Let $X=[0,1]$ and $E_x=\mathbb{C}$, $x\epsilon X$. Choose $\Lambda=\Lambda_0$ (as in 1.2) and $\Lambda'=C(\Lambda_0)=C(X)$, so that $C(\Lambda)=C(\Lambda')$ (1.17). For each $n=1,2 \ldots$ define:

$$f_n(x) = -nx+1, \quad 0\leq x\leq 1/n,$$
$$= 0 \quad , \quad 1/n\leq x\leq 1.$$

Then the set $F=\{f_n\}$ is contained in $C(X)$, so that F is an element of $E(\Lambda')$ (2.3). However, F is not equicontinuous at 0 in the classical sense, i.e. $F\notin E(\Lambda)$.

Such an example as this suggests that different Λ's also be compared in terms of their corresponding $E(\Lambda)$'s.

2.8 DEFINITION. We define $\Lambda \leqq \Lambda'$ by $E(\Lambda)\subseteq E(\Lambda')$ and $\Lambda \approx \Lambda'$ by $E(\Lambda) = E(\Lambda')$. Clearly, the relation $\leqq$ is a partial ordering and $\approx$ an equivalence relation on the continuity structures in ΠE_x. We will call the relation $\approx$ strong equivalence.

2.9 REMARKS. It follows from 2.6 and 2.7 that $\sim$ is a strictly weaker relation than $\approx$. We shall see in section 5 that $\approx$ is in turn strictly weaker than equality. Example 2.7 also shows that, in general, $C(\Lambda) \not\approx \Lambda$, although $C(\Lambda)\sim\Lambda$ and $\Lambda\leq C(\Lambda)$. Thus, roughly speaking, Λ and $C(\Lambda)$ yield the same continuity structures but possibly different equicontinuity structures.

2.10 LEMMA. Suppose $\Lambda \epsilon E(\Lambda')$. Then for each subset D of X and x_0 in D, we have:

If the subset F of $\prod_D E_x$ is Λ-equicontinuous at x_0, then F is Λ'-equicontinuous at x_0.

In particular, $E(D,\Lambda) \subseteq E(D,\Lambda')$, for all $D \subseteq X$.

Proof. Suppose f is Λ-equicontinuous at x_0 and let $\varepsilon>0$. Then there exists a neighborhood N' of x_0 in X such that for each f in F, there exists h_f in Λ, for which

$$\|f(x)-h_f(x)\| < \varepsilon/2 , \qquad x\varepsilon N'\cap D.$$

Since the set $H=\{h_f : f\varepsilon F\}$ is contained in Λ, we have $H\varepsilon E(\Lambda')$ (2.5). Hence, there exists a neighborhood $N\subseteq N'$ of x_0 in X such that for each f in F, there exists k_f in Λ' for which

$$\|h_f(x)-k_f(x)\| < \varepsilon/2 , \qquad x\varepsilon N\cap D.$$

Therefore,

$$\|f(x)-k_f(x)\| < \varepsilon , \qquad x\varepsilon N\cap D,$$

so that F is Λ'-equicontinuous at x_0.

Combining 2.3, 2.8 and 2.10, we obtain:

2.11 PROPOSITION. The following are equivalent:

(i) $\Lambda \approx \Lambda'$.

(ii) $\Lambda \varepsilon E(\Lambda')$ and $\Lambda' \varepsilon E(\Lambda)$.

(iii) For each subset D of X and x_0 in D, a family F contained in $\prod_D E_x$ is Λ-equicontinuous at x_0 if and only if it is Λ'-equicontinuous at x_0.

(iv) For each subset D of X, $E(D,\Lambda) = E(D,\Lambda')$.

In order to obtain an analogue of 1.24 for strong equivalence, we need to restate the condition on F in 1.24.

2.12 LEMMA. Let F be a subset of $C(\Lambda)\cap C(\Lambda')$. The following are equivalent:

(i) For each x_0 in X, the set $\{f(x_0):f\varepsilon F\}$ is dense in E_{x_0}.

(ii) For each x_0 in X and $\varepsilon>0$, given h in $\Lambda\cup\Lambda'$, there exists f_h in F such that $\|h(x_0)-f_h(x_0)\|<\varepsilon$.

2.13 PROPOSITION. The continuity structures Λ and Λ' are strongly equivalent if and only if there exists a set F in $E(\Lambda)\cap E(\Lambda')$ having the property that for each x_0 in X and $\varepsilon>0$, given H contained in $\Lambda\cup\Lambda'$, there exists a neighborhood N of x_0 in X satisfying: For each h in H, there exists f_h in F such that $\|f_h(x)-h(x)\|<\varepsilon$, $x\varepsilon N$.

Proof. If $\Lambda \overset{s}{\approx} \Lambda'$, choose $F=C(\Lambda)=C(\Lambda')$ and note that $\Lambda\cup\Lambda'\subseteq F$. Conversely, let G be an element of $E(\Lambda')$, x_0 an element of X and $\varepsilon>0$. Then there exists a neighborhood N'' of x_0 in X satisfying: For each g in G, there exists h_g in Λ' such that $\|g(x)-h_g(x)\|<\varepsilon/3$, for x in N''. Then $H=\{h_g:g\varepsilon G\}$ is a subset of Λ' and by hypothesis, there exists a neighborhood $N'\subseteq N''$ of x_0 in X satisfying: For each g in G, there exists f_g in F such that $\|h_g(x)-f_g(x)\|<\varepsilon/3$, $x\varepsilon N'$. Finally, since $\{f_g:g\varepsilon G\}$ is an element of $E(\Lambda)$ (2.5), there exists a neighborhood $N\subseteq N'$ of x_0 in X satisfying: For each g in G, there exists k_g in Λ such that $\|f_g(x)-k_g(x)\|<\varepsilon/3$, $x\varepsilon N$. Therefore, for $x\varepsilon N$, we have $\|g(x)-k_g(x)\|<\varepsilon$, which implies G is in $E(\Lambda)$. The other inclusion is proved symmetrically.

3. MEASURABILITY

Now let μ be a positive Radon measure on X. The notion of measurable vector field will be developed along the lines of Bourbaki [1, Chap.IV] as in Godement [1, Chap.III].

3.1 DEFINITION. Let f be a vector field defined on X. Then f is <u>(Λ,μ)-measurable</u> if for each $\varepsilon>0$ and compact subset Q of X, there exists a compact subset Q_ε of Q such that $\mu(Q-Q_\varepsilon)<\varepsilon$ and $f|Q_\varepsilon$ is Λ-continuous, i.e. $f|Q_\varepsilon$ is $\Lambda|Q_\varepsilon$-continuous. Let $M(\Lambda,\mu)$ denote the space of such vector fields. Since μ as well as Λ will be fixed for the remainder of this monograph, when there is no danger of confusion, we will delete the references to Λ and μ in the terminology. Clearly, $C(\Lambda)$ is contained in $M(\Lambda,\mu)$, for each choice of μ.

Suppose D is a locally compact subset of X. Then $\mu|D$ is also a positive Radon measure on D (the induced measure; see Bourbaki [1, p.186]) and $\Lambda|D$ is a continuity structure in $\prod_D E_x$. Denote $M(\Lambda|D,\mu|D)$ by $M(D,\Lambda,\mu)$ and recall that $M(D,\mu|D)$ is a ring. The following lemmas are easily verified.

3.2 LEMMA. The space $M(D,\Lambda,\mu)$ is an $M(D,\mu|D)$-module.

3.3 LEMMA. If f is an element of $M(D,\Lambda,\mu)$, then the scalar function $\|f\|$ is $\mu|D$-measurable, i.e. it belongs to $M(D,\mu|D)$.

Proof. This follows from 1.7.

3.4 LEMMA. If D' is a locally compact subset of X contained in D and f is a vector field on D, then $f|D'$ is measurable if f is, i.e. $f\in M(D,\Lambda,\mu)$ implies $f|D' \in M(D',\Lambda,\mu)$.

3.5 PROPOSITION (Egoroff). Suppose D is a subset of X such that X-D is locally null. Let $\{f_n\}$ be a sequence of measurable vector fields on X which converge pointwise to some vector field f on D. Then:

(i) For each compact subset Q of X and $\varepsilon>0$, there exists a compact subset Q_ε of Q such that $\mu(Q-Q_\varepsilon)<\varepsilon$ and $f_n|Q_\varepsilon$ is continuous for all n.

(ii) The sequence $\{f_n\}$ converges uniformly to f on Q_ε, i.e. $\|f_n-f\|_{Q_\varepsilon}\to 0$.

(iii) If f is extended arbitrarily to all of X, then f is measurable.

Proof. This is proved exactly as in Bourbaki [1, p.175].

3.6 LEMMA. If Λ and Λ' are two equivalent continuity structures in ΠE_x, then $M(D,\Lambda,\mu)=M(D,\Lambda',\mu)$, for each locally compact subset D of X.

Proof. See 1.23.

4. PRODUCT SPACES

Our objectives in Chapters III and IV require that we develop the previous concepts on product spaces as well. Although we begin our discussion here in general terms, we will quickly restrict our attention to the setting of future interest.

In addition to our existing notation, let Y be a locally compact Hausdorff space and ν a measure on Y. Suppose $\{E_{x,y}\}$ is a field of Banach spaces over $X \times Y$ and Γ is a continuity structure in $\Pi E_{x,y}$. Fix y in Y and define:

$$\Gamma_y = \{h(\cdot,y) : h \in \Gamma\}.$$

Since $X_y = X \times \{y\}$ is a closed subspace of $X \times Y$ (homeomorphic to X), it follows from the discussion in section 1 that Γ_y is a continuity structure in $\Pi_X E_{x,y}$. Similarly, for fixed x in X, we may define

$$_x\Gamma = \{h(x,\cdot) : h \in \Gamma\}$$

and obtain a continuity structure relative to the field $\{E_{x,y} : y \in Y\}$ of Banach spaces over $_xY = \{x\} \times Y$.

Roughly speaking, a Γ-continuous vector field splits into a family of continuous fields on one factor indexed by the other. Our problem with this setting is that it is not clear that essentially the same thing happens for vector fields on $X \times Y$ which are $(\Gamma, \mu \otimes \nu)$-measurable.

4.1 DEFINITION. We say that Γ is <u>ν-measurably restrictive for μ</u> if the following occurs: For each f in $M(\Gamma, \mu \otimes \nu)$, there exists a locally μ-null subset A of X (depending on f) such that for each x in X-A, the field $f(x,\cdot)$ on $_xY$ belongs to $M(_x\Gamma, \nu)$. In a similar manner, we define what it means for Γ to be <u>μ-measurably restrictive for ν</u>.

4.2 EXAMPLE. Suppose $E_{x,y}=E$, for all (x,y) in $X\times Y$ and ν is moderated (Bourbaki [2, p.4]). Then the family Γ_0 of constant functions from $X\times Y$ into E is ν-measurably restrictive for any μ. In this case, $M(\Gamma_0,\mu\otimes\nu)=M(X\times Y,\mu\otimes\nu,E)$ and this claim is precisely Proposition 2b of Bourbaki [2, p.89].

It is not clear under what conditions this desirable property holds more generally. In any event, our needs in the remainder of this manuscript are much more specific than is our current discussion. In fact, all we really need to do is "lift" Λ from X to $X\times Y$. Accordingly, let $\lambda:X\times Y\to X$ and $\rho:X\times Y\to Y$ denote the canonical projections. To obtain a field $\{E_{x,y}\}$ of Banach spaces on $X\times Y$, define $E_{x,y} = E_x$, for each (x,y) in $X\times Y$. Then the composition $f\to f\circ\lambda$ defines a one-to-one mapping of $\prod_X E_x$ into $\prod_{X\times Y} E_{x,y}$. In particular, denote $\{h\circ\lambda : h\varepsilon\Lambda\}$ by $\Lambda\lambda$. Thus, $\Lambda\lambda\subseteq\Pi E_{x,y}$ and $\Lambda\lambda$ is in one-to-one correspondence with Λ. The following lemmas are easily verified.

4.3 LEMMA. $\Lambda\lambda$ is a continuity structure in $\Pi E_{x,y}$.

4.4 LEMMA. Let D be a subset of X and x_0 an element of D. If the vector field f in $\prod_D E_x$ is Λ-continuous at x_0, then the field $f\lambda$ in $\prod_{D\times Y} E_{x,y}$ is $\Lambda\lambda$-continuous at (x_0,y), for all $y\varepsilon Y$. Thus, f in $C(D,\Lambda)$ implies $f\lambda$ is in $C(D\times Y,\Lambda\lambda)$. In particular, $\{f\circ\lambda : f\varepsilon C(\Lambda)\} \subseteq C(\Lambda\lambda)$.

4.5 LEMMA. Let D be a locally compact subspace of X. If f is in $M(D,\Lambda,\mu)$, then $f\lambda$ is in $M(D\times Y,\Lambda\lambda,\mu\otimes\nu)$. Therefore, $\{f\circ\lambda : f\varepsilon M(\Lambda,\mu)\}\subseteq M(\Lambda\lambda,\mu\otimes\nu)$.

4.6 LEMMA. If Λ is countably dense, separable or locally separable, then the same is true for $\Lambda\lambda$.

We next show that the correspondence $\Lambda\to\Lambda\lambda$ preserves equivalences. Let Λ and Λ' be two continuity structures in ΠE_x.

4.7 LEMMA. Let D be a subset of $X\times Y$ and (x_0,y_0) an element of D. If $C(x_0,\lambda(D),\Lambda)$ is contained in $C(x_0,\lambda(D),\Lambda')$, then $C((x_0,y_0),D,\Lambda\lambda)$ is contained in $C((x_0,y_0),D,\Lambda'\lambda)$. Consequently, $\Lambda \leq \Lambda'$ implies $\Lambda\lambda \leq \Lambda'\lambda$. In particular, $\Lambda\lambda \sim \Lambda'\lambda$ if $\Lambda \sim \Lambda'$.

Proof. Apply 1.15 - 1.19.

4.8 LEMMA. If $E(x_0,\lambda(D),\Lambda) \subseteq E(x_0,\lambda(D),\Lambda')$, then $E((x_0,y_0),D,\Lambda\lambda) \subseteq E((x_0,y_0),D,\Lambda'\lambda)$. Consequently, $\Lambda \lesssim \Lambda'$ implies $\Lambda\lambda \lesssim \Lambda'\lambda$. In particular, $\Lambda\lambda \approx \Lambda'\lambda$ if $\Lambda \approx \Lambda'$.

Proof. Apply the results of section 2.

4.9 EXAMPLE. Suppose $E_x=E$, for all x in X, and $\Lambda=\Lambda_0$ as in 1.2. Then $\Lambda_0\lambda$ is the family of constant functions from $X \times Y$ into E, i.e. "$\Lambda_0\lambda=\Gamma_0$". Hence, if Λ is any continuity structure in ΠE_x such that $C(\Lambda)=C(X,E)$, then $C(\Lambda\lambda) = C(X\times Y,E)$ by 4.7. Moreover, if $\Lambda=C(X,E)$, then $\Lambda\lambda$ is the set of f in $C(X\times Y,E)$ which are constant on the fibers ${}_xY$, for x in X.

4.10 QUESTIONS. (1) Under what conditions is $\Lambda\lambda$ ν-measurably restrictive for μ? (2) More specifically, when is Λ such that for each f in $M(\Lambda\lambda,\mu\otimes\nu)$, the function $f(x,\cdot):Y\to E_x$ is ν-measurable for locally all x in X (relative to μ)? (3) Is this true if ν is moderated?

II. CONTINUOUS SUMS OF BANACH SPACES

In this chapter, our objective is to study spaces of continuous and measurable vector fields. In the continuous case, we extend to this setting the familiar results concerning spaces of continuous functions which are either bounded or of compact support or vanish at infinity. In the measurable case, we construct the vector field analogues of L^p-spaces. These are the L^p-continuous sums of the Banach spaces making up the underlying field. Moreover, we also develop the vector field version of "mixed-norm" $L^{p,q}$-spaces. Finally, we discuss the duality between L^p-spaces.

5. SPACES OF CONTINUOUS VECTOR FIELDS

Given a continuity structure Λ in ΠE_x, define:

$$C_b(\Lambda) = \{f \in C(\Lambda) : \|f\|_X < \infty\} ,$$

$$C_o(\Lambda) = \{f \in C(\Lambda) : \|f\| \text{ vanishes at } \infty \text{ on } X\} ;$$

and

$$C_c(\Lambda) = \{f \in C(\Lambda) : \text{supp}(f) \text{ is compact}\} .$$

Clearly, we have the following chain of inclusions:

$$C_c(\Lambda) \subseteq C_o(\Lambda) \subseteq C_b(\Lambda) \subseteq C(\Lambda) ,$$

where $C(\Lambda)$ is a vector space and a $C(X)$-module. Our next concern is to establish the vector field analogues of the usual properties of the three subspaces and the relationships between them.

5.1 PROPOSITION. (i) The space $C_b(\Lambda)$ is a Banach space relative to the norm $\|\cdot\|_X$. (ii) It is also a $C_b(X)$-module. (iii) For each f in $C_b(\Lambda)$, the scalar function $\|f\|$ belongs to $C_b(X)$.

Proof. Part (iii) follows from 1.7. It is clear that $C_b(\Lambda)$ is a normed linear space relative to $\|\cdot\|_X$, so that (ii) follows from 1.9. Thus, to complete the proof of (i), it remains to show that $C_b(\Lambda)$ is complete. This is done as in the classical case (using 1.12).

5.2 PROPOSITION. (i) The space $C_o(\Lambda)$ is a Banach space relative to the norm $\|\cdot\|_X$. (ii) This space is also a $C_b(X)$-module. (iii) For each f in $C_o(\Lambda)$, the scalar function $\|f\|$ belongs to $C_o(X)$.

Proof. This proposition follows essentially from the previous one.

Before proceeding to $C_c(\Lambda)$, we consider a very interesting and important example of $C_o(\Lambda)$.

5.3 EXAMPLE. Let A be a C*-algebra with primitive ideal space denoted by X (Dixmier [1, §3]). Each x in X is a closed ideal in A. Let π_x denote the canonical projection of A onto $A/x=A_x$. Then $\{A_x\}$ is a field of C*-algebras over X which is locally compact but not Hausdorff (in general) relative to the hull-kernel topology. For each a in A, define:

$$h_a(x) = \pi_x(a) , \qquad x\varepsilon X.$$

Then h_a is an element of ΠA_x and $\Lambda = \{h_a : a\varepsilon A\}$ is a subspace of ΠA_x having the property that

$$\{h_a(x) : a\varepsilon A\} = A_x ,$$

for all x in X. Unfortunately, the scalar functions $\|h_a\|$ are not continuous in general; however, they are if X is Hausdorff (Dixmier [1,§10]). Note that this is a comparatively strong property for a C*-algebra to have, although it occurs in many cases. For example, all abelian and, more interestingly, some non-abelian C*-algebras have it. In any event, suppose X is Hausdorff so that Λ is a continuity structure in ΠA_x. Then $C_o(\Lambda)$ can be given a canonical C*-algebra structure making it isomorphic to A. If X is not Hausdorff, certain modifications can be made in the above construction to obtain similar but less complete results (see Fell [1]).

5.4 PROPOSITION. (i) The space $C_c(\Lambda)$ is a dense subspace of $C_o(\Lambda)$. (ii) This space is also a $C(X)$-module. (iii) For each f in $C_c(\Lambda)$, the scalar function $\|f\|$ is in $C_c(X)$. (iv) For each x in X, the space $\{f(x) : f \in C_c(\Lambda)\}$ is equal to E_x.

Proof. Parts (ii) and (iii) are clear and part (i) is a consequence of the complete regularity of X. Part (iv) follows from 1.16.

The next result says that $C_c(X)\cdot\Lambda$ is locally uniformly dense in $C_c(\Lambda)$.

5.5 LEMMA. Let f be an element of $C_c(\Lambda)$ and Q a compact subset of X. Given $\varepsilon>0$, there exists g in $C_c(X)\cdot\Lambda$ such that $\|f-g\|_Q<\varepsilon$.

Proof. Apply 1.11.

As consequences of the previous three propositions, we see that $C_b(\Lambda)$, $C_o(\Lambda)$ and $C_c(\Lambda)$ are also continuity structures in ΠE_x. Hence it is of interest to know how they compare with each other in terms of our two equivalence relations. For convenience, temporarily let $\Lambda_1=C_c(\Lambda)$, $\Lambda_2=C_o(\Lambda)$, $\Lambda_3=C_b(\Lambda)$, and $\Lambda_4=C(\Lambda)$.

5.6 PROPOSITION. The continuity structures $\Lambda_i, i=1,2,3,4$ are strongly equivalent to each other, i.e. $\Lambda_i \approx \Lambda_j$, $i,j=1,2,3,4$.

Proof. Since $\Lambda_1 \subseteq \Lambda_2 \subseteq \Lambda_3 \subseteq \Lambda_4$, we have $\Lambda_i \in E(\Lambda_{i+1})$, $i=1,2,3$, by 2.3. Thus, $E(\Lambda_i) \subseteq E(\Lambda_{i+1})$, $i=1,2,3$ (2.10). On the other hand, if F is an element of $E(\Lambda_4)$, then F is contained in $C(\Lambda_4)$ which is the same as $C(\Lambda)$ (1.17). Conversely, if F is a subset of $C(\Lambda)=\Lambda_4$, then F is an element of $E(\Lambda_4)$ (2.3). Therefore,

$$E(\Lambda_4) = \{F \subseteq \Pi E_x : F \subseteq C(\Lambda)\}.$$

However, if F is a subset of $C(\Lambda)$, then it is easily seen that F is an element of $E(\Lambda_1)$. Hence $E(\Lambda_4) \subseteq E(\Lambda_1)$, which completes the proof.

5.7 REMARK. In view of this proposition, it is worth recalling that Λ is equivalent (1.22) (but not strongly equivalent (2.9)) to $C(\Lambda)$. Consequently, we see that in general, $\Lambda \sim \Lambda_i$, but $\Lambda \not\approx \Lambda_i$, $i=1,2,3,4$.

6. LEBESGUE SPACES

We next construct the familiar spaces of p-summable vector fields along the lines of Bourbaki [1, Chap. IV] as in Godement [1, Chap. III].

Suppose μ is a measure on X. For each f in ΠE_x and p in $[1,\infty)$ define:

$$\|f\|_p = (\int^{-}\|f(x)\|^p dx)^{1/p}$$

and let

$$\mathcal{F}^p(\mu) = \{f \in \Pi E_x : \|f\|_p < \infty\}.$$

6.1 LEMMA. (i) $\mathcal{F}^p(\mu)$ is a complete vector space relative to the semi-norm $\|\cdot\|_p$. (ii) The space $C_c(\Lambda)$ is contained in $\mathcal{F}^p(\mu)$ and $\|\cdot\|_p$ is a norm on $C_c(\Lambda)$.

Therefore, as usual, we may define $\mathcal{L}^p(\Lambda,\mu)$ to be the closure of $C_c(\Lambda)$ in $\mathcal{F}^p(\mu)$.

6.2 THEOREM. Let g be a vector field defined on X. Then g belongs to $\mathcal{L}^p(\Lambda,\mu)$ if and only if g is measurable (i.e. $g \in M(\Lambda,\mu)$) and $\|g\|_p < \infty$. Hence,

$$\mathcal{L}^p(\Lambda,\mu) = \mathcal{F}^p(\mu) \cap M(\Lambda,\mu).$$

Next we define:

$$\begin{aligned} \mathcal{N}(\mu) &= \{g \in \Pi E_x : g=0, \mu\text{-a.e.}\} \\ &= \{g \in \Pi E_x : \|g\|_p = 0\} , \end{aligned}$$

which is independent of p. Then $\mathcal{N}(\mu)$ is a closed subspace of $\mathcal{L}^p(\Lambda,\mu)$ and we may define $L^p(\Lambda,\mu)$ to be $\mathcal{L}^p(\Lambda,\mu)/\mathcal{N}(\mu)$. This is a Banach space with norm also denoted by $\|\cdot\|_p$. In particular, if X is discrete and μ is counting measure, then $L^1(\Lambda,\mu)$ is just the direct sum of the Banach spaces E_x, x in X. If, in addition, the E_x are Hilbert spaces, then $L^2(\Lambda,\mu)$ is the Hilbert space direct sum of the E_x.

6.3 EXAMPLE. Suppose E is an arbitrary Banach space and $\omega: X \to (0,\infty)$ is a

μ-measurable locally bounded function. Let E_x be the Banach space obtained by renorming E by $\|\cdot\|_x$, where

$$\|v\|_x = \omega(x)^{-1}\|v\| , \qquad v\varepsilon E.$$

Likewise, define

$$\Lambda = \{h_v : v\varepsilon E\} ,$$

where

$$h_v(x) = \omega(x)^{-1}v , \qquad x\varepsilon X.$$

Then Λ is a continuity structure in ΠE_x, with

$$C(\Lambda) = \{\omega^{-1}\cdot f : f\varepsilon C(X,E)\}$$

and

$$L^p(\Lambda,\mu) = \{\omega^{-1}\cdot f : f\varepsilon L^p(X,\mu,E)\}.$$

6.4 REMARKS. (1) If X is a locally compact group and ω is a suitable weight function, then $L^1(\Lambda,\mu)$ is the underlying Banach space of a Beurling algebra. (2) If, in particular, ω is a "norm function" (Fell [2, p.73]), then there is a correspondence between certain Banach space representations of G and $L^1(\Lambda,\mu)$.

Before constructing $L^\infty(\Lambda,\mu)$, we consider the problem of singling out certain useful dense subspaces of $L^p(\Lambda,\mu)$. Let $\mathcal{B}$ denote the family of pre-compact Borel subsets of X. If B is an element of $\mathcal{B}$ and h an element of Λ, then $\gamma_B h$ belongs to $\mathcal{L}^p(\Lambda,\mu)$ and $\|\gamma_B h\|_p \le \|h\|_p$.

6.5 PROPOSITION. The span of the set $\{\gamma_B h : B\varepsilon\mathcal{B}, h\varepsilon\Lambda\}$ in $\mathcal{L}^p(\Lambda,\mu)$ is dense.

Proof. Since $C_c(\Lambda)$ is dense in $\mathcal{L}^p(\Lambda,\mu)$, it suffices to approximate f in $C_c(\Lambda)$. Let Q denote the support of f. Suppose U is a pre-compact open (so that $\mu(U)>0$) neighborhood of Q in X. Given $\varepsilon>0$, for each x in Q, there exists a compact neighborhood Q_x of x in X and h_x in Λ such that $Q_x \subseteq U$ and

$$\|f(y)-h_x(y)\| < \frac{\varepsilon}{\mu(U)^{1/p}} , \qquad y\varepsilon Q_x.$$

Since $\{Q_x : x\varepsilon Q\}$ is a neighborhood cover of compact Q, there exist $x_1, \ldots, x_n$ in Q, such that $Q\subseteq Q'\subseteq U$, where $Q'=Q_1\cup\ldots\cup Q_n$ and $Q_i=Q_{x_i}$, $i=1,\ldots,n$. Likewise, denote h_{x_i} by h_i and define:

$$B_i = Q_i-(Q_1\cup\ldots\cup Q_{i-1}) ,$$

so that B_i is an element of $\mathcal{B}$, for each $i=1,\ldots,n$. Note also that the B_i are mutually disjoint and

$$Q'=Q_1\cup\ldots\cup Q_n = B_1\cup\ldots\cup B_n.$$

Let $g=\Sigma\gamma_i h_i$, where $\gamma_i=\gamma_{B_i}$, $i=1,\ldots,n$. Then g belongs to $\mathcal{L}^p(\Lambda,\mu)$ and

$$\begin{aligned}
\|f-g\|_p^p &= \int_X \|f(x)-g(x)\|^p dx \\
&= \int_{Q'} \|f(x)-g(x)\|^p dx \\
&= \sum_{i=1}^{n} \int_{B_i} \|f(x)-g(x)\|^p dx \\
&= \sum_{i=1}^{n} \int_{B_i} \|f(x)-h_i(x)\|^p dx \\
&< \sum_{i=1}^{n} \frac{\varepsilon^p}{\mu(U)} \mu(B_i) \\
&< \frac{\varepsilon^p \mu(Q')}{\mu(U)} \\
&< \varepsilon^p ,
\end{aligned}$$

which completes the proof.

6.6 PROPOSITION. The span $C_c(X)\cdot\Lambda$ is dense in $\mathcal{L}^p(\Lambda,\mu)$.

Proof. By the previous result, it suffices to show the following: Given B in $\mathcal{B}$, h in Λ and $\varepsilon>0$, there exists ϕ in $C_c(X)$ such that $\|\gamma_B h-\phi h\|_p<\varepsilon$. By the regularity of μ (Hewitt and Stromberg [1, 12.40]), there exists a compact subset Q of B such that

$$\mu(B-Q) < (\tfrac{\varepsilon}{b})^p ,$$

where $b = \|h\|_B$. Hence,

$$\|\gamma_B h - \gamma_Q h\|_p^P = \int_{B-Q} \|h(x)\|^P dx$$

$$< \varepsilon^P ,$$

so that we may assume B is compact. Now choose an open pre-compact subset U of X which contains B and let $a = \|h\|_U$. Again by the regularity of μ, there exists an open neighborhood V of B contained in U such that

$$\mu(V-B) < (\frac{\varepsilon}{a})^P .$$

choose ϕ in $C_c(X)$ such that $\operatorname{supp}(\phi) \subseteq V$, $0 \leq \phi \leq 1$ and $\phi(x) = 1$, $x \in B$. Then ϕh belongs to $C_c(\Lambda)$ (5.4) and

$$\|\phi h - \gamma_B h\|_p^P = \int_{V-B} \|\phi(x) - h(x)\|^P dx$$

$$\leq a^P \mu(V-B)$$

$$< \varepsilon^P ,$$

which completes the proof.

6.7 COROLLARY. If $X = X_1 \times X_2$, then the span $C_c(X_1) \otimes C_c(X_2) \bullet \Lambda$ is dense in $\mathcal{L}^P(\Lambda, \mu)$.

Proof. Recall that $C_c(X_1) \otimes C_c(X_2)$ is uniformly dense in $C_c(X_1 \times X_2)$.

Next we construct the Lebesque space $L^\infty(\Lambda, \mu)$. For f in ΠE_x, define

$$\|f\|_\infty = \inf \{\|f\|_{X-N} : N \text{ is locally } \mu\text{-null}\}$$

and let $$\mathcal{F}^\infty(\mu) = \{f \in \Pi E_x : \|f\|_\infty < \infty\}.$$

Then f belongs to $\mathcal{F}^\infty(\mu)$ if and only if there exists g in ΠE_x such that $\|g\|_X < \infty$ and $f = g$ locally μ-almost everywhere.

6.8 LEMMA. $\mathcal{F}^{\infty}(\mu)$ is a complete vector space relative to the semi norm $\|\cdot\|_{\infty}$.

Now define

$$\mathcal{L}^{\infty}(\Lambda,\mu) = M(\Lambda,\mu) \cap \mathcal{F}^{\infty}(\mu)$$

and

$$\mathcal{N}^{\infty}(\mu) = \{g \in \Pi E_x : g=0, \text{ loc. } \mu\text{-a,e.}\}$$

$$= \{g \in \mathcal{F}^{\infty}(\mu) : \|g\|_{\infty}=0\}.$$

Then $\mathcal{N}^{\infty}(\mu)$ (resp. $\mathcal{L}^{\infty}(\Lambda,\mu)$) is a closed subspace of $\mathcal{L}^{\infty}(\Lambda,\mu)$ (resp. $\mathcal{F}^{\infty}(\mu)$) and we may define

$$L^{\infty}(\Lambda,\mu) = \mathcal{L}^{\infty}(\Lambda,\mu)/\mathcal{N}^{\infty}(\mu) ,$$

which is thus a Banach space with norm also denoted by $\|\cdot\|_{\infty}$.

6.9 LEMMA. (i) $C_b(\Lambda)$ is contained in $\mathcal{L}^{\infty}(\Lambda,\mu)$ and $\|f\|_{\infty} \leq \|f\|_X$, for f in $C_b(\Lambda)$. (ii) Equality holds in (i) for all f if and only if $\text{supp}(\mu) = X$. In this case, $C_b(\Lambda)$ may be identified with a subspace of $L^{\infty}(\Lambda,\mu)$ which is proper in general.

It is well-known that the classical Lebesque space $L^p(X,\mu,E)$ is a Hilbert space if and only if $p=2$ and E is a Hilbert space (provided that the support of μ contains at least two points). It is natural to ask under what conditions the Banach space $L^p(\Lambda,\mu)$ will be a Hilbert space. Besides, in order to discuss Hilbert-Schmidt and trace-class integral operators on $L^p(\Lambda,\mu)$, we will need to know the answer to this question. We shall see that (as might be expected) the answer in the classical case extends as well as it can to the vector field setting. First we have:

6.10 THEOREM. If E_x is a Hilbert space for each x in the support of μ, then $L^2(\Lambda,\mu)$ is a Hilbert space.

Proof. The parallelogram law is satisfied on $C_c(\Lambda)$.

To resolve the converse question, we require the following:

6.11 LEMMA. Let x be in the support of μ and suppose $\psi_1, \dots, \psi_n$ are non-negative elements of $C(X)$. Then given $\delta>0$, there exists θ in $C_c(X)$ such that $\theta \geq 0$,

$$\int \theta(y)dy = 1$$

and

$$\left|\int \theta(y)\psi_i(y)dy - \psi_i(x)\right| < \delta , \qquad i=1,\dots,n.$$

Proof. For each i, let Q_i be a compact neighborhood of x for which

$$|\psi_i(x) - \psi_i(y)| < \frac{\delta}{n} , \qquad y \in Q_i .$$

Then $Q = Q_1 \cap \dots \cap Q_n$ is a compact neighborhood of x and there exists θ' in $C_c(X)$ such that $\theta'(x)=1$, $\theta' \geq 0$ and $\mathrm{supp}(\theta') \subseteq Q$. Since x is in the support of μ, we have

$$a = \int \theta'(y)dy > 0 ,$$

so we may define $\theta = (1/a)\theta'$. Then θ belongs to $C_c(X)$, the support of θ is contained in Q and

$$\int \theta(y)dy = 1$$

Also,

$$\begin{aligned} \left|\int_X \theta(y)\psi_i(y)dy - \psi_i(x)\right| &= \left|\int_X \theta(y) \, (\psi_i(y)-\psi_i(x))dy\right| \\ &\leq \int_Q \theta(y) \, |\psi_i(y)-\psi_i(x)| \, dy \\ &< \delta \qquad i=1,\dots,n. \end{aligned}$$

6.12 THEOREM. If $1 \leq p < \infty$ and $L^p(\Lambda,\mu)$ is a Hilbert space, then E_x is a Hilbert space for each x in the support of μ.

Proof. Let f_1 and f_2 be elements of $C_c(\Lambda)$ and denote f_1+f_2 and f_1-f_2 by f_3 and f_4 respectively. Also, define:

$$\psi_i(x) = \|f_i(x)\|^p, \qquad x \in X,$$

so that ψ_i is in $C_c(X)$ and $\psi_i \geq 0$, for $i=1,2,3,4$. Let $\varepsilon>0$. Since the real function $r \to r^{2/p}$ is continuous, there exists $\delta>0$ such that

$$|r^{2/p} - \psi_i(x)^2| < \frac{\varepsilon}{8},$$

whenever $|r-\psi_i(x)| < \delta$, r a real number. By 6.11, there exists θ in $C_c(X)$ such that $\theta \geq 0$,

$$\int \theta(y)dy = 1$$

and

$$|\int \theta(y)\psi_i(y)dy - \psi_i(x)| < \delta,$$

i.e.

$$| \|\theta^{1/p} f_i\|_p^p - \psi_i(x)| < \delta,$$

so that

$$| \|\theta^{1/P} f_i\|_p^2 - \psi_i(x)^{2/P}| < \frac{\varepsilon}{8}, \qquad i=1,2,3,4.$$

Moreover, the vector fields $\theta^{1/P}f_1$ and $\theta^{1/P}f_2$ belong to $C_c(\Lambda)$ (5.4) and, by hypothesis, must satisfy the parallelogram law, i.e.

$$\|\theta^{1/P}f_4\|_p^2 + \|\theta^{1/P}f_3\|_p^2 - 2\|\theta^{1/P}f_2\|_p^2 - 2\|\theta^{1/P}f_1\|_p^2 = 0$$

Consequently,

$$| \|f_1(x)-f_2(x)\|^2 + \|f_2(x)+f_1(x)\|^2 - 2\|f_2(x)\|^2 - 2\|f_2(x)\|^2 |$$

$$= |[\psi_4(x)^{2/P} + \psi_3(x)^{2/P} - 2\psi_2(x)^{2/P} - 2\psi_1(x)^{2/P}]$$

$$- [\|\theta^{1/P}f_4\|_p^2 + \|\theta^{1/P}f_3\|_p^2 - 2\|f_2\|_p^2 - 2\|f_2\|_p^2 - 2\|f_1\|_p^2]|$$

$$\leq |\psi_4(x)^{2/P} - \|\theta^{1/P}f_4\|_p^2| + |\psi_3(x)^{2/P} - \|\theta^{1/P}f_3\|_p^2|$$

$$+ 2|\psi_2(x)^{2/P} - \|\theta^{1/P}f_2\|_p^2| + 2|\psi_1(x)^{2/P} - \|\theta^{1/P}f_1\|_p^2|$$

$$< \frac{3\varepsilon}{4}, \qquad \varepsilon \text{ arbitrary.}$$

Therefore,

$$\|f_1(x)+f_2(x)\|^2 + \|f_1(x)-f_2(x)\|^2 = 2\|f_1(x)\|^2 + 2\|f_2(x)\|^2 ,$$

for all f_1 and f_2 in $C_c(\Lambda)$, which completes the proof by 5.4.

The two previous results suggest it would be beneficial to replace Λ and μ by their restrictions to the support of μ. This will be further suggested by the remaining results on the Hilbert space question for $L^p(\Lambda,\mu)$. Hence, before proceeding, we digress briefly to justify making these replacements.

Let Y denote an arbitrary closed subspace of X. Then μ induces a measure $\mu|Y$ on Y (Bourbaki [1, Chap. IV]) and a measure μ_Y on X of base μ and density function γ_Y (see Bourbaki [2, Chap. V] for the details). Define $Rf = f|Y$, for f in ΠE_X and

$$Tg(x) = g(x), \ x \varepsilon Y ,$$
$$= 0 , \ x \notin Y ,$$

for g in $\Pi_Y E_y$. As in section 7 of Bourbaki [2, Chap. V], we see that $TRf = \gamma_Y f$ and $RTg = g$. The mapping R is a linear isometry of $\mathcal{L}^p(\Lambda,\mu_Y)$ $(1 \leq p \leq \infty)$ onto $\mathcal{L}^p(\Lambda|Y,\mu|Y)$ and the mapping T is a linear isometry of $\mathcal{L}^p(\Lambda|Y,\mu|Y)$ into $\mathcal{L}^p(\Lambda,\mu_Y)$ which is onto modulo $\mathcal{N}(\mu_Y)$. Furthermore, since

$$T(\mathcal{N}(\mu|Y)) = \mathcal{N}(\mu_Y) ,$$
$$T(\mathcal{N}^\infty(\mu|Y)) = \mathcal{N}^\infty(\mu_Y) ,$$
$$R(\mathcal{N}(\mu_Y)) = \mathcal{N}(\mu|Y)$$

and

$$R(\mathcal{N}^\infty(\mu_Y)) = \mathcal{N}^\infty(\mu|Y) ,$$

it follows that T and R pass to isometric isomorphisms between $L^p(\Lambda|Y,\mu|Y)$ and $L^p(\Lambda,\mu_Y)$ which are inverses of each other. In particular, if Y is the support of μ, then $\mu_Y = \mu$ and we have that $L^p(\Lambda,\mu)$ is isometrically isomorphic to $L^p(\Lambda|Y,\mu|Y)$, where $\mu|Y$ is such that $\mathrm{supp}\,(\mu|Y) = Y$.

Therefore, in asking under what conditions $L^p(\Lambda,\mu)$ is a Hilbert space, we may assume X is the support of μ.

Let us return to our line of thinking before the digression.

6.13 PROPOSITION. If $C_o(\Lambda)$ is a Hilbert space, then E_x is a Hilbert space, for each x in X. If $L^\infty(\Lambda,\mu)$ is a Hilbert space, then E_x is a Hilbert space, for each x in the support of μ.

Proof. The first statement is proved as in 6.12. For the proof of the second statement, replace $L^\infty(\Lambda,\mu)$ by $L^\infty(\Lambda|Y,\mu|Y)$, for $Y=\mathrm{supp}(\mu)$. Then the conclusion follows from 6.9, together with the first statement applied to $\Lambda|Y$.

The next result completes the solution to the Hilbert space question for $L^p(\Lambda,\mu)$.

6.14 THEOREM. (i) Suppose the support of μ has at least two points. If $L^p(\Lambda,\mu)$ $(1\leq p\leq\infty)$ is a Hilbert space, then $p=2$. (ii) If the space X has at least two points, then $C_o(\Lambda)$ is not a Hilbert space.

Proof. (i) Assume first that $2\leq p<\infty$ and $L^p(\Lambda,\mu)$, i.e. $L^p(\Lambda|Y,\mu|Y)$ is a Hilbert space with Y the support of μ. By 6.12, each E_x, for x in Y, is a Hilbert space. Hence if q is such that $1/p + 1/q = 1$, then $1\leq q\leq 2$ and $L^q(\Lambda|Y,\mu|Y)$, i.e. $L^q(\Lambda,\mu)$ is also a Hilbert space since it is isomorphic to the dual space of $L^p(\Lambda|Y,\mu|Y)$ (Godement [1, Chap. III]). Thus, we may assume $1\leq p\leq 2$ in the finite case.

Next, let x and y be distinct points in the support of μ. Suppose A and B are disjoint compact neighborhoods of x and y respectively in X. Choose f and g in $C_c(\Lambda)$ so that $f(x) \neq o \neq g(y)$, $\mathrm{supp}(f)\subseteq A$ and $\mathrm{supp}(g)\subseteq B$ (5.4). Then $\|f\|_p \neq o \neq \|g\|_p$ and, by hypothesis, f and g satisfy the parallelogram law. Furthermore, by choice of f and g, we have:

$$\|f(z)+g(z)\|^p = \|f(z)-g(z)\|^p = \|f(z)\|^p + \|g(z)\|^p ,$$

for each z in X, so that

$$\|f+g\|_p^2 = \|f-g\|_p^2 = [\|f\|_p^p + \|g\|_p^p]^{2/p} .$$

Consequently,

$$[\|f\|_p^p + \|g\|_p^p]^{2/p} = \|f\|_p^2 + \|g\|_p^2 .$$

Letting $a = \|f\|_p^p$ and $b = \|g\|_p^p$, we conclude that $p=2$ from the following claim:

If $a \neq 0 \neq b$ and $[a+b]^{2/p} = a^{2/p}+b^{2/p}$ for $1 \leq p \leq 2$, then $p=2$.

The case $p=\infty$ in part (i) of the theorem and part (ii) are handled similarly.

6.15 REMARK. Of course, if X is a singleton $\{x\}$, then

$$E_x \cong C_o(\Lambda) \cong L^p(\Lambda,\mu) , \qquad 1 \leq p \leq \infty.$$

Thus, these are Hilbert spaces if and only if E_x is.

The following results on completely orthonormal bases are due to Godement (Godement [1, p.92]). For this discussion, suppose E_x is a Hilbert space, for each x in X.

6.16 DEFINITION. A family $\{e_\alpha\}$ of vector fields in $M(\Lambda,\mu)$ is a <u>completely orthonormal base</u> if the following conditions are satisfied:

(i) For each pair of distinct α and β, the scalar function $x \to \langle e_\alpha(x), e_\beta(x) \rangle$ is locally $\mu\otimes\mu$-negligible.

(ii) For each α, the function $x \to \|e_\alpha(x)\|$ is not locally μ-negligible and attains only the values 0 and 1.

(iii) The family $\{e_\alpha\}$ is a maximal family having properties (i) and (ii).

Condition (iii) may be replaced by the following:

(iii) If a vector field f in $M(\Lambda,\mu)$ is such that $\langle f(x), e_\alpha(x)\rangle = 0$ locally μ-almost everywhere, for each α, then f is locally μ-negligible.

6.17 LEMMA. Complete orthonormal bases exist in general.

Proof. Zorn's Lemma.

6.18 PROPOSITION. Let $\{e_\alpha\}$ be a complete orthonormal base. Let $f \in \Pi E_x$. Then f belongs to $L^2(\Lambda,\mu)$ if and only if:

(i) The functions $x \to \langle f(x), e_\alpha(x)\rangle$ are μ-measurable, all α.

(ii)
$$\Sigma \int |\langle f(x), e_\alpha(x)|^2 \, dx < \infty.$$

(iii)
$$f(x) = \sum_\alpha \langle f(x), e_\alpha(x)\rangle \, e_\alpha(x) ,$$

for μ-almost all x in X, the summation being taken over those α for which the coefficient functions are not μ-negligible.

6.18 PROPOSITION. Let $\{e_\alpha\}$ be as above and $f \in \Pi E_x$. Then f is (Λ,μ)-measurable if and only if:

(i) The functions $x \to \langle f(x), e_\alpha(x)\rangle$ are μ-measurable, all α.

(ii) For each compact subset Q of X, there exists a sequence $\{\alpha_n\}$ of indices such that

$$f(x) = \sum_n \langle f(x), e_{\alpha_n}(x)\rangle \, e_{\alpha_n}(x) ,$$

for μ-almost all x in Q.

Now suppose that Λ is countably dense and let $\{h_n\}$ be a sequence in Λ satisfying (iii) of 1.1.

6.19 PROPOSITION. If Λ is countably dense, then there exists a sequence $\{e_n\}$ which is a completely orthonormal base.

Proof. Apply the Gram-Schmidt process to $\{h_n\}$.

6.20 PROPOSITION. Suppose Λ is countably dense with $\{e_n\}$ as in 6.19. Let $f \varepsilon \Pi E_x$. The following are equivalent:

(i) f is (Λ,μ)-measurable.

(ii) The functions $x \to \langle f(x), e_n(x)\rangle$ are μ-measurable, all n.

(iii) The functions $x \to \langle f(x), h_n(x)\rangle$ are μ-measurable, all n.

7. MIXED-NORM LEBESGUE SPACES

In Section 16 we will find it necessary to consider Lebesque spaces of "mixed-norm" type. Such spaces were developed in Benedek and Panzone [1] for the case of scalar-valued functions. In this section, we construct such spaces of vector fields, paralleling the development in Section 6 as much as we can. For the sake of clarity, we restrict our attention to the case of two variables.

Suppose X and Y are locally compact Hausdorff spaces equipped with measures μ and ν respectively. As before, if there is no possibility of confusion, it will be convenient to write $dx = d\mu(x)$ and $dy = d\nu(y)$. The product measure $\mu\otimes\nu$ will be handled analogously, i.e. $d(x,y) = d\mu\otimes\nu(x,y)$. Choose p and q such that $1\leq p,q<\infty$ (we leave the infinite cases for the interested reader to pursue). Suppose further that $\{E_{x,y}\}$ is a field of Banach spaces over $X \times Y$ and Λ is a continuity structure in $\Pi E_{x,y}$. For each f in $\Pi E_{x,y}$, define:

$$\|f\|_{p,q} = (\int^{-}(\int^{-} \|f(x,y)\|^{p}dx)^{q/p}dy)^{1/q} ,$$

so that $0 \leq \|f\|_{p,q} \leq \infty$, in general. Let

$$F^{p,q}(\mu,\nu) = \{f\varepsilon\Pi E_{x,y} : \|f\|_{p,q}<\infty \text{ and } f \text{ is } \mu\otimes\nu\text{-moderated}\}$$

and

$$M^{p,q}(\Lambda,\mu,\nu) = F^{p,q}(\mu,\nu) \cap M(\Lambda,\mu\otimes\nu).$$

The reader should note that in the case of $F^{p}(\mu)$ (§6), its elements are automatically μ-moderated. However, in the mixed-norm case, it is not clear whether the elements of $F^{p,q}(\mu,\nu)$ are automatically $\mu\otimes\nu$-moderated. The crux of the problem is (1) of 7.6 at the end of this section.

7.1 PROPOSITION.

(i) $F^{p,q}(\mu,\nu)$ is a semi-normed linear space relative to $\|\cdot\|_{p,q}$.

(ii) $\mathcal{M}^{p,q}(\Lambda,\mu,\nu)$ is a complete linear subspace of $\mathcal{F}^{p,q}(\mu,\nu)$.

(iii) The space $C_c(\Lambda)$ is contained in $\mathcal{M}^{p,q}(\Lambda,\mu,\nu)$ and $\|\cdot\|_{p,q}$ is a norm on $C_c(\Lambda)$.

Proof. (i) is straightforward. (ii) and (iii) are proved in the usual ways, making use of Bourbaki [2, p.93].

7.2 PROPOSITION. The following are equivalent for f in $\mathcal{M}^{p,q}(\Lambda,\mu,\nu)$:
(i) $\|f\|_{p,q} = 0$. (ii) f is $\mu\otimes\nu$-null. (iii) For ν-almost all y in Y, the field $f(\cdot,y)$ is μ-null.

Proof. These equivalences are consequences of the results of Bourbaki [2, p.93].

Thus, if we define $\mathcal{L}^{p,q}(\Lambda,\mu,\nu)$ to be the closure of $C_c(\Lambda)$ in $\mathcal{M}^{p,q}(\Lambda,\mu,\nu)$, then $\mathcal{N}(\mu\otimes\nu)$ is the closure of 0 in $\mathcal{L}^{p,q}(\Lambda,\mu,\nu)$ and the quotient Banach space

$$L^{p,q}(\Lambda,\mu,\nu) = \mathcal{L}^{p,q}(\Lambda,\mu,\nu)/\mathcal{N}(\mu\otimes\nu)$$

is the mixed-norm Lebesgue space we desire. Note that the construction given here agrees with that in Benedek and Panzone [1] where the two overlap.

7.3 REMARKS. If p=q, then we know that the elements of $\mathcal{F}^{p,q}(\mu,\nu)$, i.e. $\mathcal{F}^{p}(\mu\otimes\nu)$, are automatically $\mu\otimes\nu$-moderated. Also, the space $\mathcal{F}^{p}(\mu\otimes\nu)$ is complete and $\mathcal{M}^{p,q}(\Lambda,\mu,\nu) = \mathcal{L}^{p,q}(\Lambda,\mu,\nu)$. It is not clear whether any of these properties holds in the case $p \neq q$ as well. The crux of this problem is (2) of 7.6.

The next result will be very useful to us in Section 16.

7.4 THEOREM. The subspace $C_c(X\times Y)\cdot\Lambda$ is dense in $\mathcal{L}^{p,q}(\Lambda,\mu,\nu)$.

Proof. It suffices to show $C_c(X\times Y)\cdot\Lambda$ is dense in $C_c(\Lambda)$. Let f be a non-zero element of $C_c(\Lambda)$ and $\varepsilon>0$. Choose compact subsets Q_1 and Q_2 of X and Y respectively such that $\mathrm{supp}(f) \subseteq Q$, for $Q=Q_1\times Q_2$. Also choose precompact open neighborhoods V_1 and V_2 of Q_1 and Q_2 in X and Y respectively and let $V = V_1 \times V_2$. By 5.5, there exists k in $C_c(X\times Y)\cdot\Lambda$ such that

$$\|f-k\|_Q < \frac{\varepsilon}{2^{1/p+1/q}\mu(Q_1)^{1/p}\mu(Q_2)^{1/q}} .$$

Furthermore, by the regularity of μ and ν, there exist open subsets U_1 and U_2 of X and Y respectively satisfying: For $c = \|f\|_V + \|k\|_V$, we have

$$Q_i \subseteq U_i \subseteq V_i , \qquad i=1,2,$$

$$\mu(Q_1-Q_1) < \frac{\varepsilon^p}{2^{1+p/q}\mu(Q_2)^{p/q}c^p} ,$$

and

$$\mu(U_2-Q_2) < \frac{\varepsilon^q}{2\ \mu(V_1)^{q/p}c^q} .$$

Now choose θ in $C_c(X\times Y)$ such that $0 \le \theta \le 1$, $\text{supp}(\theta) \subseteq U$ and $\theta(x,y) = 1$, $(x,y) \in Q$. Then the field $g = \theta k$ belongs to $C_c(X\times Y)\cdot\Lambda$, $\|f-g\|_Q = \|f-k\|_Q$, and $\|g\| < \|k\|$. To complete the proof, we next show that $\|f-g\|_{p,q} < \varepsilon$. We have

$$\begin{aligned}
\|f-g\|_{p,q} &= (\textstyle\int_X(\int_Y\|f(x,y)-g(x,y)\|^p dx)^{q/p}dy)^{1/q} \\
&= (\textstyle\int_{U_2}(\int_{U_1}\|f(x,y)-g(x,y)\|^p dx)^{q/p}dy)^{1/q} \\
&\le (\textstyle\int_{U_2-Q_2}(\int_{U_1}\|f(x,y)-g(x,y)\|^p dx)^{q/p}dy \\
&\quad + \textstyle\int_{Q_2}(\int_{U_1}\|f(x,y)-g(x,y)\|^p dx)^{q/p}dy)^{1/q} .
\end{aligned}$$

We compute the two iterated integrals separately. For the first one:

$$\begin{aligned}
\|f-g\|_U &\le (\|f\|_U + \|g\|_U) \\
&\le (\|f\|_U + \|k\|_U) \\
&\le (\|f\|_V + \|k\|_V) \\
&= c,
\end{aligned}$$

so that

$$\int_{U_2-Q_2}\left(\int_{U_1}\|f(x,y)-g(x,y)\|^p dx\right)^{q/p} dy < \frac{\varepsilon^q}{2}.$$

For the second one:

$$\int_{Q_2}\left(\int_{U_1}\|f(x,y)-g(x,y)\|^p dx\right)^{q/p} dy \le \int_{Q_2}\left(\int_{U_1-Q_1}\|f(x,y)-g(x,y)\|^p dx\right.$$

$$\left. + \int_{Q_1}\|f(x,y)-g(x,y)\|^p dx\right)^{q/p} dy$$

$$\le \int_{Q_2}\left(c^p\mu(U_1-Q_1) + \int_{Q_1}\|f(x,y)-k(x,y)\|^p dx\right)^{q/p} dy$$

$$< \frac{\varepsilon^q}{2} .$$

Consequently, $\|f-g\|_{p,q} < \varepsilon$, and the proof is complete.

7.5 COROLLARY. The subspace $C_c(X) \otimes C_c(Y)\cdot\Lambda$ is dense in $\mathcal{L}^{p,q}(\Lambda,\mu,\nu)$.

7.6 QUESTIONS. (1) If $f\varepsilon\Pi E_{x,y}$ and $\|f\|_{p,q} < \infty$, then the function

$$y \to \int^{-}\|f(x,y)\|^p dx$$

is ν-moderated (Bourbaki [2, p.6]) and finite for μ-almost all x in X. Thus, the field $x \to f(x,y)$ is μ-moderated for ν-almost all y in Y. Does this imply that f is $\mu\otimes\nu$-moderated? (2) The space $\mathcal{M}^{p,q}(\Lambda,\mu,\nu)$ was introduced because we don't know if the mixed-norm version of 6.2 is true: Is g an element of $\mathcal{L}^{p,q}(\Lambda,\mu,\nu)$ if and only if g belongs to $M(\Lambda,\mu\otimes\nu)$ and $\|g\|_{p,q}<\infty$? As in the classical case, it may be verified that the forward implication is true. Thus, the question here is really whether the conditions are sufficient for g to belong to $\mathcal{L}^{p,q}(\Lambda,\mu,\nu)$. Furthermore, in Section 6, the space $\mathcal{F}^p(\mu)$ was found to be complete. It is not clear that this is also true of $\mathcal{F}^{p,g}(\mu,\nu)$ in general. This is important since $\mathcal{L}^{p,q}(\Lambda,\mu,\nu)$ is defined to be the completion of $C_c(\Lambda)$ in $\mathcal{F}^{p,q}(\mu,\nu)$. Both of these difficulties are eliminated by introduction of the space $M^{p,q}(\Lambda,\mu,\nu)$ which is really the one in which we wish to complete $C_c(\Lambda)$.

8. DUAL CONTINUITY STRUCTURES

We continue to assume that Λ is a continuity structure in ΠE_x and μ a measure on X. The family $\{E_x^*\}$ of dual spaces is also a field of Banach spaces over X. Thus, we may consider continuity structures Λ^* in ΠE_x^*. Intuitively, a structure Λ^* should be compatible with Λ in some reasonable sense. There are two canonical choices for the sense of this compatibility--namely, "continuous" and "measurable." We will choose the former since it is the more useful and (to us) the more appropriate. Our objective in this section is to develop the properties of such compatible pairs (Λ,Λ^*) of continuity structures and recall the familiar duality relationship existing between the spaces $L^p(\Lambda,\mu)$ and $L^{p^*}(\Lambda^*,\mu)$, for $1 \leq p < \infty$ and p^* such that $1/p + 1/p^* = 1$ (Dinculeanu [1]). If f and f* belong to $\underset{D}{\Pi} E_x$ and $\underset{D}{\Pi} E_x^*$ respectively, it will be convenient to denote the scalar function $x \to \langle f(x), f^*(x)\rangle$ on D by $\langle f:f^*\rangle$.

8.1 DEFINITION. The continuity structures Λ and Λ^* in ΠE_x and ΠE_x^* respectively form a <u>duality pair</u> if for each h in Λ and each h* in Λ^*, the function $\langle h:h^*\rangle$ is continuous on X.

8.2 EXAMPLE. If $E_x = E$, for all x in X, then the spaces Λ_o and Λ_o^* of constant mappings from X into E and E* respectively form a duality pair.

8.3 EXAMPLE. If X is discrete, then $(\Pi E_x, \Pi E_x^*)$ is a duality pair.

8.4 EXAMPLE. If each E_x is a Hilbert space, for x in X, then we obtain a duality pair if we define:

$$\Lambda^* = \{h^* : h \in \Lambda\} ,$$

where h* is the element of ΠE_x^* given by

$$h^*(x)v = \langle v, h(x)\rangle , \qquad v \in E_x,\ x \in X.$$

For a given Λ, it is not clear whether there always exists a Λ^* for which (Λ,Λ^*) is a duality pair (see (1) of 8.14 below). In any event, the following properties of a given duality pair (Λ,Λ^*) are easily verified.

8.5 LEMMA. Suppose D is a subset of X and x_0 is a point in D. If f belongs to $C(x_0,D,\Lambda)$ and f^* belongs to $C(x_0,D,\Lambda^*)$, then the scalar function $<f:f^*>$ is in $C(x_0,D)$. In particular, $<f:f^*> \in C(D)$, whenever $f\in C(D,\Lambda)$ and $f^*\in C(D,\Lambda^*)$.

8.6 COROLLARY. If $f\in C(\Lambda)$ and $f^*\in C_c(\Lambda^*)$ (or $f\in C_c(\Lambda)$ and $f^*\in C(\Lambda^*)$), then $<f:f^*> \in C_c(X)$.

8.7 LEMMA. If $f\in C_b(\Lambda)$ and $f^*\in C_b(\Lambda^*)$, then $<f:f^*> \in C_b(X)$. If, in addition, $f\in C_0(\Lambda)$ or $f^*\in C_0(\Lambda^*)$, then $<f:f^*> \in C_0(X)$.

8.8 LEMMA. If $f\in M(\Lambda,\mu)$ and $f^*\in M(\Lambda^*,\mu)$, then $<f:f^*> \in M(X,\mu)$.

8.9 LEMMA. Let $\lambda,\rho,\Lambda\lambda$ and $\Lambda^*\lambda$ be as in section 4, with $X = Y$ and $\mu = \nu$. If (Λ,Λ^*) is a duality pair, then so is $(\Lambda\lambda,\Lambda^*\lambda)$.

8.10 LEMMA. If D is a locally compact subset of X and (Λ,Λ^*) is a duality pair, then so is $(\Lambda|D,\Lambda^*|D)$.

8.11 LEMMA. Suppose Λ_i is a continuity structure in ΠE_x and Λ_i^* is a continuity structure in ΠE_x^*, for $i=1,2$. If (Λ_1,Λ_1^*) is a duality pair and $\Lambda_1\sim\Lambda_2$, $\Lambda_1^*\sim\Lambda_2^*$, then (Λ_2,Λ_2^*) is a duality pair.

In Dinculeanu [1, p.5], the following duality relationships were shown to exist amongst Lebesgue spaces constructed relative to a duality pair (see also Godement [1]).

8.12 THEOREM. Let p and p^* be such that $1\le p<\infty$ and $1/p + 1/p^* = 1$. For each f^* in $L^{p^*}(\Lambda^*,\mu)$, define:

$$T_{f^*}(f) = \int<f(x),f^*(x)>dx, \qquad f\in L^p(\Lambda,\mu).$$

Then T_{f*} is a continuous linear functional on $L^p(\Lambda,\mu)$, i.e. it is an element of $L^p(\Lambda,\mu)^*$, and the mapping $T : L^{p^*}(\Lambda^*,\mu) \to L^p(\Lambda,\mu)^*$ is an into linear isometry.

Therefore, the space $L^{p^*}(\Lambda^*,\mu)$ may be viewed as a closed subspace of $L^p(\Lambda,\mu)^*$ via the mapping T. However, it is not clear if the subspace is the whole space in general. It is the whole space under certain additional assumptions.

8.13 THEOREM. Suppose also that at least one of the following is true in 8.12:

(i) Λ and Λ^* are countably dense.

(ii) For each x in the support of μ, the space E_x is a Hilbert space.

Then T is onto, i.e. $L^{p^*}(\Lambda^*,\mu)$ is isometrically isomorphic to $L^p(\Lambda,\mu)^*$ via the mapping T.

8.14 QUESTIONS.

(1) Given the continuity structure Λ, consider the set

$$\{h^* \in \Pi E_x^* : \langle h:h^* \rangle \in C(X)\}.$$

Is this set a continuity structure? If so then, together with Λ, it forms a duality pair.

(2) If (Λ,Λ^*) is a duality pair and Λ is countably dense, separable or locally separable, does Λ^* have the same property?

(3) In 8.12, is the space

$$\{T_{f*} : f^* \in L^{p^*}(\Lambda^*,\mu)\}$$

a norming subspace of $L^p(\Lambda,\mu)^*$ (assuming it's not the whole space)? More generally, is

$$\{T_{f*} : f^* \in C_c(\Lambda^*)\}$$

a norming subspace of $L^p(\Lambda,\mu)^*$?

(4) In 8.13, does it suffice to assume that Λ and Λ^* are separable or locally separable? Note that 8.12 and 8.13 were proved under the assumption that (Λ,Λ^*) was just a "measurable" duality pair, i.e. $\langle h:h^*\rangle \in M(X,\mu)$, for $h\in\Lambda$, $h^*\in\Lambda^*$.

III. KERNEL FIELDS

Needless to say, it is the existence of a kernel which makes an operator an integral operator. In this chapter, we introduce the concept of kernel in the vector field context and study it from the points of view of smoothness and type. The first part of the chapter develops the three familiar types of continuity and measurability for a kernel--namely, weak, strong, and uniform. The second part focuses on two special types of kernel--namely, those whose values are either finite-rank or Hilbert-Schmidt operators.

9. NORM FORMULAS

Let X and Y be locally compact Hausdorff spaces. Given fields $\{E_x\}$, $\{F_y\}$ of Banach spaces over X,Y, we observe that $\{\mathrm{Hom}(F_y,E_x)\}$ is then a field of Banach spaces over $X\times Y$, where $\mathrm{Hom}(F_y,E_x)$ is chosen to be the space over (x,y) in $X\times Y$.

9.1 DEFINITION. If D is a subset of $X\times Y$, then a <u>kernel</u> K on D is an element of $\Pi_D \mathrm{Hom}(F_y,E_x)$. Thus, for each (x,y) in D, $K(x,y)$ is a bounded linear operator of F_y into E_x. If $D = X\times Y$, then we will say simply that K is a kernel. If K is a kernel on D and f is a vector field in $\Pi_{\rho(D)} F_y$, then the vector field $(x,y) \to K(x,y)f(y)$ belongs to $\Pi_D E_{x,y}$, where $E_{x,y} = E_x$, $(x,y)\in D$ (recall §4). For convenience, we will denote this field by f_K.

The purpose of this section is to establish two formulas for $\|K(x,y)\|$ -- a "strong" one and a "weak" one--which will be useful in the next two sections. Let Λ,Δ be continuity structures in $\Pi E_x, \Pi F_y$.

9.2 PROPOSITION. If K is a kernel, then

$$\|K(x,y)\| = \sup_{\substack{f\in C_c(\Delta) \\ \|f\|_Y\le 1}} \|K(x,y)f(y)\| , \qquad (x,y)\in X\times Y$$

Proof. It is clear that $\|K(x,y)\|$ is not less than the quantity on the right. For the other inequality, let v in F_y be such that $\|v\| = 1$. Then there exists g_v in $C(\Delta)$ such that $g_v(y) = v$ (1.16); thus, $\|g_v\|_Y \geq 1$. Since the function $\|g_v\|$ is continuous, given $\varepsilon>0$, there exists a precompact neighborhood N of y in Y for which

$$| \, \|g_v(t)\| - \|g_v(y)\| \, | < \varepsilon ,$$

i.e.

$$| \, \|g_v(t)\| - 1 \, | < \varepsilon ,$$

so that

$$\|g_v(t)\| < 1 + \varepsilon , \qquad t\in N.$$

Let Q be a compact neighborhood of y in Y contained in N and choose ϕ in $C_c(Y)$ satisfying: $0\le\phi\le 1$, $\mathrm{supp}(\phi)\subseteq N$ and $\phi(t)=1$, $t\in Q$. Then $f_v = \phi g_v$ belongs to $C_c(\Delta)$, $f_v(t) = g_v(t)$, $t\in Q$ and

$$\begin{aligned}\|f_v\|_Y &= \sup_{t\in N} \|f_v(t)\| \\ &\le \sup_{t\in N} \|g_v(t)\| \\ &\le 1 + \varepsilon .\end{aligned}$$

Hence, for all v in F_y of norm 1, we have:

$$\|K(x,y)v\| = \|K(x,y)f(y)\| ,$$

for some f in $C_c(\Delta)$ such that $\|f\|_Y < 1 + \varepsilon$. Consequently,

$$\begin{aligned}\|K(x,y)\| &= \sup_{\substack{v\varepsilon F_y\\ \|v\|\le 1}} \|K(x,y)v\| \\ &\le \sup_{\substack{f\varepsilon C_c(\Delta)\\ \|f\|_Y\le 1+\varepsilon}} \|K(x,y)f(y)\| \\ &= \sup_{\substack{f\varepsilon C_c(\Delta)\\ \|f\|_Y\le 1}} \|K(x,y)f(y)\| ,\end{aligned}$$

since ε is arbitrary.

Now suppose (Λ,Λ^*) and (Δ,Δ^*) are duality pairs as in section 8 for $(\Pi E_x,\Pi E_x^*)$ and $(\Pi F_y,\Pi F_y^*)$ respectively.

9.3 PROPOSITION. If K is a kernel, then

$$\|K(x,y)\| = \sup_{\substack{f\varepsilon C_c(\Delta)\\ \|f\|_Y\le 1}} \ \sup_{\substack{g^*\varepsilon C_c(\Lambda^*)\\ \|g^*\|_X\le 1}} |\langle K(x,y)f(y),g^*(x)\rangle| , \quad (x,y)\varepsilon X\times Y .$$

Proof. Recall that for v in E_x , we have:

$$\|v\| = \sup_{\substack{v^*\varepsilon E_x^*\\ \|v^*\| = 1}} |\langle v,v^*\rangle| .$$

Suppose $\varepsilon>0$ and let v,v^* be elements of E_x,E_x^*. As in the proof of 9.2, there exists g^* in $C_c(\Lambda^*)$ such that $g^*(x) = v^*$ and $\|g^*\|_X \le 1+\varepsilon$, Hence, for v^* such that $\|v^*\| = 1$, we have:

$$\langle v,v^*\rangle = \langle v,g^*(x)\rangle ,$$

so that

$$\|v\| \leq \sup_{\substack{g^* \in C_c(\Lambda^*) \\ \|g^*\|_X \leq 1+\varepsilon}} |\langle v, g^*(x)\rangle| .$$

Since ε is arbitrary, this implies that

$$\|v\| \leq \sup_{\substack{g^* \in C_c(\Lambda^*) \\ \|g^*\|_X \leq 1}} |\langle v, g^*(x)\rangle| .$$

On the other hand, if g^* in $C_c(\Lambda^*)$ is such that $\|g^*\|_X \leq 1$, then

$$|\langle v, g^*(x)\rangle| \leq \|v\| ,$$

so that

$$\sup_{\substack{g^* \in C_c(\Lambda^*) \\ \|g^*\|_X \leq 1}} |\langle v, g^*(x)\rangle| \leq \|v\| .$$

Consequently,

$$\|v\| = \sup_{\substack{g^* \in C_c(\Lambda^*) \\ \|g^*\|_X \leq 1}} |\langle v, g^*(x)\rangle| , \qquad v \in E_X .$$

The proof is then completed by applying the previous proposition.

10. STRONG CONTINUITY AND MEASURABILITY

This section is concerned with defining and studying the notions of strong continuity and measurability for a kernel. In addition to the notation of section 9, let D be a subset of $X \times Y$ and (x_o, y_o) a point in D.

10.1 DEFINITION. If K is a kernel on D, then K is <u>strongly (Λ,Δ)-continuous</u> at (x_o,y_o) if for each h in Δ, the vector field h_K defined on D is $\Lambda\lambda$-continuous at (x_o,y_o), i.e. if h_K belongs to $C((x_o,y_o),D,\Lambda\lambda)$ (recall §4). Denote the space of such kernels by $\mathcal{C}_s((x_o,y_o),D,\Lambda,\Delta)$. It will be convenient to introduce the following additional notation:

$$\mathcal{C}_s^b((x_o,y_o),D,\Lambda,\Delta) = \{K \varepsilon \mathcal{C}_s((x_o,y_o),D,\Lambda,\Delta) : K \text{ is locally bound at } (x_o,y_o)\}$$

$$\mathcal{C}_s(D,\Lambda,\Delta) = \bigcap_{(x_o,y_o)\varepsilon D} \mathcal{C}_s((x_o,y_o),D,\Lambda,\Delta)$$

$$\mathcal{C}_s^b(D,\Lambda,\Delta) = \bigcap_{(x_o,y_o)\varepsilon D} \mathcal{C}_s^b((x_o,y_o),D,\Lambda,\Delta)$$

If $D=X\times Y$, we will delete it. Also, if $\Lambda = \Delta$, we will write Λ for Λ,Δ. If there is no possibility of confusion, we will delete the reference to Λ,Δ in the above terminology (but not in the notation).

The following lemmas are easily verified.

10.2 LEMMA. The space $\mathcal{C}_s(D,\Lambda,\Delta)$ is a $C(D)$-module and a vector space.

10.3 LEMMA. Strong continuity is preserved by restriction. Specifically, if $D' \subseteq D$ and $K \varepsilon \mathcal{C}_s(D,\Lambda,\Delta)$, then $K|D' \varepsilon \mathcal{C}_s(D',\Lambda,\Delta)$.

10.4 LEMMA. Suppose Λ' and Δ' are also continuity structures in ΠE_x, and ΠF_y respectively. If $\Lambda \leq \Lambda'$ and $\Delta \leq \Delta'$, then

$$\mathcal{C}_s^b((x_o,y_o),D,\Lambda,\Delta) \subseteq \mathcal{C}_s^b((x_o,y_o),D,\Lambda',\Delta').$$

Equality holds if $\Lambda \sim \Lambda'$ and $\Delta \sim \Delta'$.

The following results show that the correspondence $f \to f_K$ preserves continuity for strongly continuous, locally bounded K.

10.5 LEMMA. If $K \in C_s^b((x_0,y_0),D,\Lambda,\Delta)$ and $f \in C(y_0,\rho(D),\Delta)$, then $f_K \in C((x_0,y_0),D,\Lambda\lambda)$. Thus, $f_K \in C(D,\Lambda\lambda)$, whenever $K \in C_s^b(D,\Lambda,\Delta)$ and $f \in C(\rho(D),\Delta)$.

10.6 PROPOSITION. Suppose K is strongly continuous, i.e. $K \in C_s(\Lambda,\Delta)$. Then the following are equivalent:

(i) For each f in $C(\Delta)$, the field f_K belongs to $C(\Lambda\lambda)$.

(ii) The scalar function $\|K\|$ is lower semi-continuous.

(iii) The scalar function $\|K\|$ is locally bounded, i.e. $K \in C_s^b(\Lambda,\Delta)$.

Proof. We have: (ii) implies (iii) for any function, (iii) implies (i) by 10.5, and (i) implies (ii) by 9.2.

Our next concern is the definition of strong measurability. There are two reasonable choices here--the first a bit more canonical than the second. Suppose now that D is a locally compact subspace of $X \times Y$ and μ,ν measures on X,Y. Recall that $\mu \otimes \nu | D$ is also a measure on D.

10.7 DEFINITION. Let K be a kernel on D. Then K is <u>strongly (Λ,Δ,μ,ν)-measurable</u> if for each h in Δ, the field h_K on D is $(\Lambda\lambda,\mu\otimes\nu)$-measurable, i.e. $h_K \in M(D,\Lambda\lambda,\mu\otimes\nu)$. Denote such K (resp. locally bounded K) by $M_s(D,\Lambda,\Delta,\mu,\nu)$ (resp. $M_s^b(D,\Lambda,\Delta,\mu,\nu)$).

10.8 DEFINITION. Let K be a kernel on D. Then K is <u>ultra-strongly (Λ,Δ,μ,ν)-measurable</u> if for each compact subset Q of D and $\varepsilon > 0$, there exists a compact subset Q_ε of Q for which $\mu\otimes\nu(Q-Q_\varepsilon) < \varepsilon$ and $K|Q_\varepsilon$ is strongly (Λ,Δ)-continuous, i.e. $K|Q_\varepsilon$ belongs to $C_s(Q_\varepsilon,\Lambda,\Delta)$. Denote such K (resp. locally bounded K) by $M_{us}(D,\Lambda,\Delta,\mu,\nu)$ (resp. $M_{us}^b(D,\Lambda,\Delta,\mu,\nu)$).

10.9 LEMMA. We have the following inclusions in general:

$$C_s(D,\Lambda,\Delta) \subseteq M_{us}(D,\Lambda,\Delta,\mu,\nu) \subseteq M_s(D,\Lambda,\Delta,\mu,\nu).$$

As for when the two notions of strong measurability are the same, we have the following partial response:

10.10 PROPOSITION. Suppose $\rho(D)$ is locally compact and $\Lambda|\rho(D)$ is countably dense. Then

$$M_s^b(D,\Lambda,\Delta,\mu,\nu) \subseteq M_{us}^b(D,\Lambda,\Delta,\mu,\nu).$$

Proof. Suppose K is in $M_s^b(D,\Lambda,\Delta,\mu,\nu)$. As in Godement [1, p.96], let $\{h_n\}$ be a sequence in Λ making $\Delta|\rho(D)$ countably dense. By hypothesis, we have:

$$\{(h_n)_K\} \subseteq M(D,\Lambda\lambda,\mu\otimes\nu).$$

If Q is a compact subset of D and $\varepsilon>0$, then (as in the proof of Egoroff's Theorem) there exists a compact subset Q_ε of Q such that $\mu\otimes\nu(Q-Q_\varepsilon)<\varepsilon$ and $(h_n)_K|Q_\varepsilon$ is $\Lambda\lambda$-continuous, for all n. In order to show that $K|Q_\varepsilon$ is (Λ,Δ)-strongly continuous, note that each h in Δ is a uniform-on-compacta limit of a sequence $\{f_m\}$ in $C(\Delta)$, where each f_m is a finite sum of the form $\Sigma\phi_{n,m}h_n$ (1.11). Hence,

$$(f_m)_K = \Sigma\phi_{n,m}(h_n)_K$$

and by 1.9, we have that $(f_m)_K|Q_\varepsilon$ is $\Lambda\lambda$-continuous, for each m. Since K is locally bounded, it follows that $h_K|Q_\varepsilon$ is a uniform limit on Q_ε of the fields $(f_n)_K|Q_\varepsilon$, i.e. $h_K|Q_\varepsilon$ is $\Lambda\lambda$-continuous. Therefore, $K|Q_\varepsilon$ is strongly (Λ,Δ)-continuous and the proof is complete.

10.11 LEMMA. The spaces $M_{us}(D,\Lambda,\Delta,\mu,\nu)$ and $M_s(D,\Lambda,\Delta,\mu,\nu)$ are vector spaces and $M(D,\mu\otimes\nu)$-modules. Similarly for $M_{us}^b(D,\Lambda,\Delta,\mu,\nu)$ and $M_s^b(D,\Lambda,\Delta,\mu,\nu)$ relative to $M^b(D,\mu\otimes\nu)$.

The next result shows that the correspondence $f \to f_K$ preserves measurability for strongly measurable K. Suppose, in addition, that $\rho(D)$ is locally compact.

10.12 PROPOSITION. Let K be an element of $M_s(D,\Lambda,\Delta,\mu,\nu)$. Then for each f in $M(\rho(D),\Delta,\nu)$, the field f_K belongs to $M(D,\Lambda\lambda,\mu\otimes\nu)$. (Note that here K need not be locally bounded as was the case with strong continuity.)

Proof. (Adapted from Godement [1, P.95]) Let f be an element of $M(\rho(D),\Delta,\nu)$, Q a compact subset of D and $\varepsilon>0$. Of course, $\lambda(Q)$ and $\rho(Q)$ are also compact; for convenience, denote $\rho(Q)$ by A. Let b be a positive number satisfying $\mu(\lambda(Q)) \leq b$. By the measurability of f, there exists a compact subset A_ε of A such that $\nu(A-A_\varepsilon) < \varepsilon/2b$ and $f|A_\varepsilon$ is Δ-continuous. Thus, $f|A_\varepsilon$ is a uniform limit of a sequence $\{f_n\}$ of vector fields in $\prod_{A_\varepsilon} \Xi_y$ of the form

$$\sum_i \phi_{n,i}\, h_{n,i} \mid A_\varepsilon \qquad \text{(finite sum)},$$

where $\phi_{n,i} \in C(A_\varepsilon)$ and $h_{n,i} \in \Delta$, for all n,i (1.11). Let

$$Q'_\varepsilon = \{(x,y) \in Q : y \in A_\varepsilon\}$$
$$= Q \cap (\lambda(Q)\times A_\varepsilon).$$

Then for each integer n and (x,y) in Q'_ε, we have:

$$(f_n)_K(x,y) = \sum_i (\phi_{n,i}\rho)(x,y)(h_{n,i})_K(x,y) ,$$

i.e.

$$(f_n)_K = \sum_i (\phi_{n,i}\rho)(h_{n,i})_K$$

on Q'_ε, for all n. Since K is strongly measurable, it follows that

$$(h_{n,i})_K \in M(D,\Lambda\lambda,\mu\otimes\nu), \qquad \text{all } n,i.$$

Consequently,

$$(h_{n,i})_K | Q'_\varepsilon \ \varepsilon \ M(Q'_\varepsilon, \Lambda\lambda, \mu\otimes\nu).$$

Hence,

$$(\phi_{n,i}\rho)(h_{n,i})_K | Q'_\varepsilon \ \varepsilon \ M(Q'_\varepsilon, \Lambda\lambda, \mu\otimes\nu) \ ,$$

since $\phi_{n,i}\rho \ \varepsilon \ C(X\times A_\varepsilon)$ and $Q'_\varepsilon \subseteq X\times A_\varepsilon$, so that

$$(f_n)_K | Q'_\varepsilon \ \varepsilon \ M(Q'_\varepsilon, \Lambda\lambda, \mu\otimes\nu) \qquad (1.9).$$

Now for (x,y) in Q'_ε , we have y in A_ε and $f_n(y)\to f(y)$ (uniformly). Thus,

$$(f_n)_K(x,y) \to (f_K)(x,y)$$

and f_K belongs to $M(Q'_\varepsilon, \Lambda\lambda, \mu\otimes\nu)$ (3.5). It follows that there exists a compact subset Q_ε of Q'_ε for which $\mu\otimes\nu(Q_\varepsilon - Q'_\varepsilon) < \varepsilon/2$ and $f_K | Q_\varepsilon$ belongs to $C(Q_\varepsilon, \Lambda\lambda)$, where

$$Q_\varepsilon - Q'_\varepsilon \subseteq \lambda(Q)\times(A - A_\varepsilon).$$

Therefore,

$$\begin{aligned} \mu\otimes\nu(Q - Q_\varepsilon) &\leq \mu\otimes\nu(Q - Q'_\varepsilon) + \mu\otimes\nu(Q'_\varepsilon - Q_\varepsilon) \\ &\leq \nu(A - A_\varepsilon)\mu(\lambda(Q)) + \varepsilon/2 \\ &< \varepsilon \ , \end{aligned}$$

so that $f_K \ \varepsilon \ M(D, \Lambda\lambda, \mu\otimes\nu)$. This completes the proof.

11. WEAK CONTINUITY AND MEASURABILITY

Next we define and study the concepts of weak continuity and weak measurability. For the purposes of this section, suppose Λ^* and Δ are continuity structures in ΠE_x^* and ΠF_y respectively. Once again, let D be a subset of $X\times Y$ and (x_0,y_0) a point in D.

11.1 DEFINITION. Suppose K is a kernel on D. Then K is <u>weakly (Λ^*,Δ)-continuous</u> at (x_0,y_0) if for each h in Δ and k^* in Λ^*, the scalar function

$$(x,y) \to \langle K(x,y)h(y),k^*(x)\rangle$$

on D (denoted by $\langle h_K:k^*\lambda\rangle$) is continuous at (x_0,y_0). Denote the space of such kernels by $\mathcal{C}_w((x_0,y_0),D,\Lambda^*,\Delta)$. Also let:

$$\mathcal{C}_w^b(x_0,y_0),D,\Lambda^*,\Delta) = \{K\varepsilon\mathcal{C}_w((x_0,y_0),D,\Lambda^*,\Delta):K \text{ is locally bounded at } (x_0,y_0)\}$$

$$\mathcal{C}_w(D,\Lambda^*,\Delta) = \bigcap_{(x,y)\varepsilon D}\mathcal{C}_w((x,y),D,\Lambda^*,\Delta) ,$$

$$\mathcal{C}_w^b(D,\Lambda^*,\Delta) = \bigcap_{(x,y)\varepsilon D}\mathcal{C}_w^b((x,y),D,\Lambda^*,\Delta) .$$

As before, if $D = X\times Y$, we will delete it. When there is no possibility of confusion, we'll also delete the reference to (Λ^*,Δ) in the terminology only.

The following lemmas are easily verified:

11.2 LEMMA. Strong continuity implies weak continuity. Specifically, for each (x_0,y_0) and D as above, we have:

$$\mathcal{C}_s((x_0,y_0),D,\Lambda,\Delta) \subseteq \mathcal{C}_w((x_0,y_0),D,\Lambda^*,\Delta) ,$$

where Λ is such that (Λ,Λ^*) is a duality pair.

11.3 LEMMA. The space $\mathcal{C}_w(D,\Lambda^*,\Delta)$ is a $C(D)$-module and a vector space.

11.4 LEMMA. Weak continuity is preserved by restriction. Specifically, if $D' \subsetneq D$ and $K \varepsilon \mathcal{C}_w(D,\Lambda^*,\Delta)$, then $K|D' \varepsilon \mathcal{C}_w(D',\Lambda^*,\Delta)$.

11.5 LEMMA. Suppose $\Lambda^{*\prime}$ and Δ' are also continuity structures in ΠE_x^* and ΠF_y respectively. If $\Lambda^* \leq \Lambda^{*\prime}$ and $\Delta \leq \Delta'$, then

$$\mathcal{C}_w^b((x_0,y_0),D,\Lambda^*,\Delta) \subsetneq \mathcal{C}_w^b((x_0,y_0),D,\Lambda^{*\prime},\Delta').$$

Equality holds if $\Lambda^* \sim \Lambda^{*\prime}$ and $\Delta' \sim \Delta$.

The following results show that the correspondence

$$(f,g^*) \to \langle f_K:g^*\lambda\rangle$$

preserves continuity for weakly continuous, locally bounded K.

11.6 LEMMA. Let K be an element of $\mathcal{C}_w^b((x_0,y_0),D,\Lambda^*,\Delta)$. If f is in $C(y_0,\rho(D),\Delta)$ and g^* is in $C(x_0,\lambda(D),\Lambda^*)$, then $\langle f_K:g^*\lambda\rangle$ belongs to $C((x_0,y_0),D)$. Thus, $\langle f_K:g^*\lambda\rangle$ is in $C(D)$, whenever $K\varepsilon\mathcal{C}_w^b(D,\Lambda^*,\Delta)$, $f\varepsilon C(\rho(D),\Delta)$ and $g^*\varepsilon C(\lambda(D),\Lambda^*)$.

11.7 PROPOSITION. Let K be an element of $\mathcal{C}_w(\Lambda^*,\Delta)$. The following are equivalent:

(i) For each f in $C(\Delta)$ and g^* in $C(\Lambda^*)$, the function $\langle f_K:g^*\lambda\rangle$ is continuous.

(ii) The function $\|K\|$ is lower semi-continuous on $X\times Y$.

(iii) The function $\|K\|$ is locally bounded on $X\times Y$, i.e. $K \varepsilon \mathcal{C}_w^b(\Lambda^*,\Delta)$.

Proof. We have: (i) implies (ii) by 9.3, (ii) implies (iii) as in 10.6, and (iii) implies (i) by 11.6.

Next we consider the notion of weak measurability. As in the case of strong measurability, there are two reasonable definitions for this notion--one a bit more canonical than the other. Once again, assume D is a locally compact subspace of $X\times Y$.

11.8 DEFINITION. Suppose K is a kernel on D. Then K is <u>weakly $(\Lambda^*,\Delta,\mu,\nu)$-measurable</u> if for each h in Δ and k^* in Λ^*, the scalar function $\langle h_K : k^*\lambda\rangle$ on D is $\mu\otimes\nu$-measurable, i.e. belongs to $M(D,\mu\otimes\nu)$. Denote such K by $\mathcal{M}_w(D,\Lambda^*,\Delta,\mu,\nu)$ and the locally bounded such K by $\mathcal{M}_w^b(D,\Lambda^*,\Delta,\mu,\nu)$.

11.9 DEFINITION. Suppose K is a kernel on D. Then K is <u>ultra-weakly $(\Lambda^*,\Delta,\mu,\nu)$-measurable</u> if for each compact subset Q of D, and $\varepsilon>0$, there exists a compact subset Q_ε of Q such that $\mu\otimes\nu(Q-Q_\varepsilon)<\varepsilon$ and $K|Q_\varepsilon$ is weakly (Λ^*,Δ)-continuous. Denote these K by $\mathcal{M}_{uw}(D,\Lambda^*,\Delta,\mu,\nu)$ and the locally bounded ones by $\mathcal{M}_{uw}^b(D,\Lambda^*,\Delta,\mu,\nu)$.

11.10 LEMMA. We have the following inclusions:

$$\mathcal{C}_w(D,\Lambda^*,\Delta) \subseteq \mathcal{M}_{uw}(D,\Lambda^*,\Delta,\mu,\nu) \subseteq \mathcal{M}_w(D,\Lambda^*,\Delta,\mu,\nu) ,$$

$$\mathcal{M}_{us}(D,\Lambda,\Delta,\mu,\nu) \subseteq \mathcal{M}_{uw}(D,\Lambda^*,\Delta,\mu,\nu)$$

and

$$\mathcal{M}_s(D,\Lambda,\Delta,\mu,\nu) \subseteq \mathcal{M}_w(D,\Lambda^*,\Delta,\mu,\nu).$$

As for when the notions of weak measurability are equivalent, we have the following partial response:

11.11 LEMMA. Suppose $\lambda(D)$ and $\rho(D)$ are locally compact and $\Delta|\rho(D)$ and $\Lambda^*|\lambda(D)$ are countably dense. Then

$$\mathcal{M}_w^b(D,\Lambda^*,\Delta,\mu,\nu) \subseteq \mathcal{M}_{uw}^b(D,\Lambda^*,\Delta,\mu,\nu).$$

Proof. The proof here is the "scalar" version of that given in 10.10.

11.12 LEMMA. The spaces $\mathcal{M}_{uw}(D,\Lambda^*,\Delta,\mu,\nu)$ and $\mathcal{M}_w(D,\Lambda^*,\Delta,\mu,\nu)$ are $M(D,\mu\otimes\nu)$-modules and vector spaces. Similarly for the locally bounded cases.

The next result shows that the correspondence $f \to f_K$ preserves measurability for a weakly measurable kernel K. Suppose now that D is a locally compact subspace of $X\times Y$ such that $\lambda(D)$ and $\rho(D)$ are also locally compact subspaces of X and Y respectively.

11.13 PROPOSITION. Let K be an element of $M_w(D,\Lambda^*,\Delta,\mu,\nu)$. Then $<f_K:g^*\lambda>$ is in $M(D,\mu\otimes\nu)$ for each f in $M(\rho(D),\Delta,\nu)$ and g* in $M(\lambda(D),\Lambda^*,\nu)$ (compare with 11.6).

Proof. The proof here is the "scalar" version of that given in 10.12.

12. UNIFORM CONTINUITY AND MEASURABILITY

The results of sections 10 and 11 show two things: (1) The extension of weak and strong continuity to the context of vector fields is canonical in that (roughly speaking) the canonical continuity structures are replaced by the given ones in the definitions. (2) The resulting developments are parallels of each other to a great extent. However, in the case of uniform continuity, the situation appears (at first) to be quite different in both regards. For Banach space-valued functions, uniform continuity is the same as norm continuity, i.e. continuity relative to a continuity structure consisting of constant functions. If this approach is taken here, then the desired parallelism between uniform, strong and weak continuity will be lost. On the other hand, in order to be faithful to the generalization of norm-continuity given in section 1, there should be an underlying continuity structure upon which norm-continuity is defined. Thus, to accommodate these opposing motivations, we find it appropriate to develop two notions of continuity here--uniform continuity and norm-continuity. These will agree in the classical case and in the discrete case. Moreover, they will agree (as best they can) in general (12.6).

For the purpose of norm-continuity, suppose $\{B_{x,y}\}$ is a field of Banach spaces over $X \times Y$, where each $B_{x,y}$ is a closed subspace of $\mathrm{Hom}(F_y, E_x)$. Note that we have purposely chosen the more general condition that $B_{x,y}$ be contained in (rather than equal to) $\mathrm{Hom}(F_y, E_x)$ because of (iii) of 1.1. For example, we may wish to consider only the compact operators in $\mathrm{Hom}(F_y, E_x)$, in which case a potential continuity structure in $\Pi\mathrm{Hom}(F_y, E_x)$ may fail to satisfy (iii) of 1.1 (see also (3) of 12.2 below). There is another way to remedy this difficulty which is notationally more convenient--namely, by means of a "pre-continuity structure." Specifically, a subset Ω of $\Pi\mathrm{Hom}(F_y, E_x)$ will be called a <u>pre-continuity structure</u> if it satisfies (i) and (ii) of 1.1. Of course, Ω becomes an actual continuity structure if each fiber $\mathrm{Hom}(F_y,E_x)$ is replaced by the obvious closed subspace. For us, norm-continuity will mean continuity relative to such a structure Ω which is also compatible with Λ and Δ in the following sense: Each element of Ω should be "inherently uniformly continuous" relative to Λ and Δ. In order to determine the specific nature of this compatibility, we first consider the classical case. The next result characterizes classical norm-continuity for a kernel function in terms which are meaningful for kernel fields.

12.1 PROPOSITION. Suppose $E_x=E$ and $F_y=F$, for all $x\in X$ and $y\in Y$, and Ω_0 is the space of constant mappings from $X\times Y$ into $\mathrm{Hom}(F,E)$ (with Λ_0 and Δ_0 as in 1.2). Let D be a subset of $X\times Y$ and $(x_0,y_0)\in D$. A kernel K on D is continuous at (x_0,y_0) (relative to Ω_0) if and only if for each bounded subset V of F, the set of functions $\{v_K: v\in V\}$ from D into E, given by

$$v_K(x,y) = K(x,y)v\,, \qquad (x,y)\in D\,,$$

is equicontinuous at (x_0,y_0) (relative to $\Lambda_0\lambda$).

Proof. Straightforward.

Before proceeding further, some comments and observations are in order.

12.2 REMARKS. (1) The set V corresponds to a subset of Δ_0. (2) The continuity structure Ω_0 consists of continuous kernels, i.e. each element of Ω_0 satisfies the condition of the proposition. (3) For constant vector fields local uniform boundedness and uniform boundedness are the same. Note that a subset G of $\prod_D F_y$ $(D\subseteq Y)$ is locally uniformly bounded at y_0 if there exists $0<b<\infty$ and a neighborhood N of y_0 in Y such that $\|g(y)\|\le b$, for all y in $N\cap D$ and g in G. (4) If B_K is any closed subspace of $\mathrm{Hom}(F,E)$ containing the range $K(D)$ of K, then K is Ω_0-continuous at (x_0,y_0) if and only if it is Ω_K-continuous at (x_0,y_0), where Ω_K is the space of constant functions from $X \times Y$ into B_K.

First we define uniform continuity (and then norm-continuity) as motivated by the previous results and comments. Let Λ and Δ be continuity structures in ΠE_x and ΠF_y respectively.

12.3 DEFINITION. A kernel K on D is <u>uniformly (Λ,Δ)-continuous</u> at (x_0,y_0) if for each subset H of Δ which is locally uniformly bounded at y_0, the set H_K of vector fields $\{h_K : h \in H\}$ is $\Lambda\lambda$-equicontinuous at (x_0,y_0). (Observe that by 2.3, H is automatically Δ-equicontinuous at y_0.) Denote these kernels by $\mathcal{C}_u((x_0,y_0),D,\Lambda,\Delta)$ and let

$$\mathcal{C}_u(D,\Lambda,\Delta) = \bigcap_{(x,y)\in D} \mathcal{C}_u((x,y),D,\Lambda,\Delta).$$

As before, $\mathcal{C}_u(\Lambda,\Delta) = \mathcal{C}_u(X \times Y,\Lambda,\Delta)$ and the superscript ^{b} denotes local boundedness.

12.4 LEMMA. Uniform continuity implies strong continuity. Specifically,

$$\mathcal{C}_u((x_0,y_0),D,\Lambda,\Delta) \subseteq \mathcal{C}_s((x_0,y_0),D,\Lambda,\Delta),$$

for $D \subseteq X \times Y$ and $(x_0,y_0) \in D$.

12.5 DEFINITION. A pre-continuity structure Ω in $\Pi \mathrm{Hom}(F_y,E_x)$ is a <u>uniform (Λ,Δ)-pre-continuity structure</u> if $\Omega \subseteq \mathcal{C}_u(\Lambda,\Delta)$. Of course, Ω_0 above (12.1) is a uniform (Λ_0,Δ_0)-continuity structure.

12.6 PROPOSITION. Norm-continuity implies uniform continuity. Specifically, if Ω is as in 12.5, then

$$C((x_0,y_0),D,\Omega) \subseteq \mathcal{C}_u((x_0,y_0),D,\Lambda,\Delta),$$

for $D \subseteq X \times Y$ and $(x_0,y_0) \in D$. In particular, $C(\Omega) \subseteq \mathcal{C}_u(\Lambda,\Delta)$.

Proof. Let K be an element of $C((x_0,y_0),D,\Omega)$. Suppose H is a subset of Δ which is locally uniformly bounded by $b>0$ in a neighborhood N of y_0.

Given $\varepsilon>0$, there exists J in Ω and a neighborhood U' of (x_0,y_0) in $X\times Y$ such that $U' \subseteq X\times N$ and

$$\|K(x,y) - J(x,y)\| < \varepsilon/2b , \qquad (x,y) \in U'\cap D.$$

By hypothesis, H_J is $\Lambda\lambda$-equicontinuous at (x_0,y_0). Thus, there exists a neighborhood U of (x_0,y_0) in $X\times Y$ such that $U \subseteq U'$ and U satisfies: For each h in H, there exists g_h in Λ for which

$$\|J(x,y)h(y) - g_h(x)\| < \varepsilon/2 , \qquad (x,y) \in U.$$

Note that $\|h(y)\| \leq b$, for $y\varepsilon N$ and $h\varepsilon H$. Hence, for $(x,y) \in U\cap D$, we have $y\varepsilon N$ so that:

$$\|K(x,y)h(y)-g_h(x)\| \leq \|K(x,y)h(y) - J(x,y)h(y)\| + \|J(x,y)h(y) - g_h(x)\|$$
$$< \varepsilon , \qquad h\varepsilon H.$$

Therefore, H_k is $\Lambda\lambda$-equicontinuous at (x_0,y_0) and K is contained in $C_u((x_0,y_0),D,\Lambda,\Delta)$.

The following lemmas are easily verified.

12.7 LEMMA. The space $C_u(D,\Lambda,\Delta)$ is a $C(D)$-module and a vector space.

12.8 LEMMA. Restriction preserves uniform continuity. Specifically, if $D' \subseteq D$ and $K\varepsilon C_u(D,\Lambda,\Delta)$, then $K|D'\varepsilon C_u(D'\Lambda,\Delta)$.

12.9 LEMMA. Suppose Λ' and Δ' are also continuity structures in ΠE_x and ΠF_y respectively. If $\Lambda \lesssim \Lambda'$ and $\Delta \lesssim \Delta'$ (recall 2.8), then

$$C_u^b((x_0,y_0),D,\Lambda,\Delta) \subseteq C_u^b((x_0,y_0),D,\Lambda',\Delta').$$

Equality holds if $\Lambda \approx \Lambda'$ and $\Delta \approx \Delta'$.

12.10 EXAMPLE. Suppose $E_x = E$ and $F_y = F$, for all x in X and y in Y. Suppose also that Λ and Δ are arbitrary continuity structures in ΠE_x and ΠF_y respectively, with Λ_0, Δ_0 and Ω_0 as above. If $\Lambda \approx \Lambda_0$ and $\Delta \approx \Delta_0$, then $C_u^b(\Lambda,\Delta) = C_u^b(\Lambda_o,\Delta_o)$ by 12.9 and $C(\Omega_o) \subseteq C_u(\Lambda_o,\Delta_o)$ by 12.6. In fact, it is actually the case that $C(\Omega_o) = C_u(\Lambda_o,\Delta_o)$ by the following argument. Suppose K is an element of $C_u(\Lambda_o,\Delta_o)$ and (x_o,y_o) is a point of $X\times Y$. For y in Y,

$$F = \{h(y) : h\varepsilon\Delta_o\},$$

so that

$$\{h(y) : h\varepsilon\Delta_o, \|h\|_Y \leq 1\}$$

is the unit ball of F. Let H denote this set. Then H is locally uniformly bounded at y_o. By hypothesis, H_K is then $\Lambda_o\lambda$-equicontinuous at (x_o,y_o). But $\Lambda_o\lambda$ is the space of constant functions for $X\times Y$ into E (4.9). Hence, H_K is a family of mappings from $X\times Y$ into E which is equicontinuous in the usual sense. From this it follows that given $\varepsilon>0$, there exists a neighborhood U of (x_o,y_o) in $X\times Y$ such that

$$\|K(x,y)h(y) - K(x_o,y_o)h(y_o)\| < \varepsilon, \qquad (x,y)\varepsilon U,\ h\varepsilon H.$$

Since each h in H is a constant function, we have

$$\|(K(x,y) - K(x_o,y_o))h(y_o)\| < \varepsilon, \qquad (x,y)\varepsilon U,\ h\varepsilon H,$$

which implies

$$\|K(x,y) - K(x_o,y_o)\| < \varepsilon, \qquad (x,y)\varepsilon U.$$

Therefore, K is continuous and locally bounded at (x_o,y_o). We thus have:

$$C_u(\Lambda_o,\Delta_o) = C_u^b(\Lambda_o,\Delta_o) = C(\Omega_o)$$

and

$$C_u^b(\Lambda,\Delta) = C(\Omega_o).$$

12.11 EXAMPLE. If X and Y are discrete, then $\Lambda = \Pi E_x$, $\Delta = \Pi F_y$ and $\Omega = \Pi\mathrm{Hom}(F_y,E_x)$ are the essentially unique continuity structures existing (1.5). In this case

$$C(\Omega) = C_u(\Lambda,\Delta) = \Omega \qquad (12.6).$$

12.12 REMARK. The previous examples show that in the classical and discrete cases, the notions of norm-continuity and uniform continuity are the same. We do not know if this is the case in general. (See (4) of 12.19 below.)

12.13 EXAMPLE. The following question is suggested by 12.9: If $\Lambda'\sim\Lambda$ and $\Delta'\sim\Delta$, is $C_u^b(\Lambda',\Delta') = C_u^b(\Lambda,\Delta)$? The answer is "no" in general as we shall now show. Let $X=Y=[0,1]$ and $F_x=F_y=\mathbb{C}$, for $x,y\in X$. Define:

$$\begin{aligned} f_n(x) &= -nx+1, & 0 \le x \le 1/n, & \\ &= 0 & 1/n \le x \le 1, & \qquad n=1,2,\ldots . \end{aligned}$$

Then $\{f_n\}$ is a sequence in $C(X) = C(\Delta_0)$. Now let $K(x,y) = 1$, $(x,y) \in X\times X$, so that $K\in\Omega_0 \subseteq C(\Omega_0) = C_u(\Lambda_0,\Delta_0)$. Choose $\Lambda'=\Delta'=C(X)$, so that $C(\Lambda')=C(\Delta')=C(X)$ (1.17), i.e. $\Lambda'\sim\Lambda_0$ and $\Delta'\sim\Delta_0$. Our claim will be justified if we show that K does not belong to $C_u(\Lambda',\Delta')$. But this is basically a consequence of the fact that the sequence $\{f_n\}$ does not converge to the zero function on X.

The following results show that the correspondence $G\to G_K$ preserves local uniform boundedness and uniform continuity for uniformly continuous, locally bounded K.

12.14 LEMMA. Let $D\subseteq X\times Y$ and $(x_0,y_0)\in D$ and suppose K is a kernel in $C_u^b((x_0,y_0),D,\Lambda,\Delta)$. If G is a subset of $C(y_0,\rho(D),\Delta)$ which is locally uniformly bounded and Δ-equicontinuous at y_0, then G_K is $\Lambda\lambda$-equicontinuous at (x_0,y_0).

Proof. Since K is locally bounded at (x_0,y_0), there exists a neighborhood U' of (x_0,y_0) in $X\times Y$ and $a>0$ such that

$$\|K(x,y)\| \le a, \qquad (x,y)\in U'\cap D.$$

Let G be as the lemma and $\varepsilon>0$. By hypothesis, there exists a neighborhood N of y_0 in Y and $b>0$ such that:

$$\text{(i)} \qquad \|f(y)\| \le b, \qquad y\in N\cap\rho(D), f\in G.$$

(ii) For each f in G, there exists h_f in Δ for which

$$\|f(y)-h_f(y)\| < \varepsilon/2a, \qquad y\varepsilon N\cap\rho(D).$$

Let

$$H = \{h_f : f\varepsilon G\},$$

so that H is a subset of Δ, i.e. it is Δ-equicontinuous (2.3). Now for each f in G and y in $N\cap\rho(D)$, we have:

$$\|h_f(y)\| \leqq \|h_f(y) - f(y)\| + \|f(y)\|$$

$$< \varepsilon/2a + b,$$

i.e. H is locally uniformly bounded at y_0. Consequently, by hypothesis, H_K is $\Lambda\lambda$-equicontinuous at (x_0,y_0). Thus, there exists a neighborhood U of (x_0,y_0) in $X\times Y$ which is contained in $U'\cap(X\times N)$ and satisfies: For each f in G, there exists g_f in Δ such that

$$\|K(x,y)h_f(y) - g_f(x)\| < \varepsilon/2, \qquad (x,y)\varepsilon U\cap D.$$

For (x,y) in $U\cap D$, we have y in $N\cap\rho(D)$ and therefore:

$$\|K(x,y)f(y) - g_f(x)\| \leq \|K(x,y)f(y) - K(x,y)h_f(y)\| + \|K(x,y)h(y) - g_f(x)\|$$

$$\leq \|K(x,y)\| \, \|f(y)-h_f(y)\| + \|K(x,y)h(y) - g_f(x)\|$$

$$< \varepsilon, \qquad f\varepsilon G.$$

Hence, G_K is $\Lambda\lambda$-equicontinuous at (x_0,y_0).

12.15 PROPOSITION. Let K be an element of $C_u(\Lambda,\Delta)$. The following are equivalent:

(i) For each subset G of $C(\Delta)$ which is locally uniformly bounded and Δ-equicontinuous on Y, the set G_K is $\Lambda\lambda$-equicontinuous on $X\times Y$.

(ii) The scalar function $\|K\|$ is lower semi-continuous on $X\times Y$.

(iii) The scalar function $\|K\|$ is locally bounded on $X\times Y$, i.e. $K\varepsilon C_u^b(\Lambda,\Delta)$.

Proof. We have: (i) implies (ii) by 10.6 and 12.4, (ii) implies (iii) as before, and (iii) implies (i) by 12.12.

Now let μ,ν be measures on X,Y. The notion of norm-measurability relative to $(\Omega,\mu\otimes\nu)$ is then an example of section 1.3. Thus, we need discuss only uniform (Λ,Δ,μ,ν)-measurability. Once again, let D be a locally compact subspace of $X\times Y$.

12.16 DEFINITION. Let K be a kernel defined on D. Then K is <u>uniformly (Λ,Δ,μ,ν)-measurable</u> if given $\varepsilon>0$ and a compact subset Q of D, there exists a compact subset Q_ε of Q such that $\mu\otimes\nu(Q-Q_\varepsilon)<\varepsilon$ and $K|Q_\varepsilon$ is uniformly (Λ,Δ)-continuous. Denote the space of such K by $M_u(D,\Lambda,\Delta,\mu,\nu)$. As before, D will be deleted if $D=X\times Y$ and the superscript ^{b} will signify that the elements are locally bounded.

The following lemmas are easily verified.

12.17 LEMMA. We have the following inclusions:

$$C_u(D,\Lambda,\Delta) \subseteq M_u(D,\Lambda,\Delta,\mu,\nu) \subseteq M_{us}(D,\Lambda,\Delta,\mu,\nu).$$

12.18 LEMMA. If Ω is a uniform (Λ,Δ)-pre-continuity structure in $\Pi\mathrm{Hom}(F_y,E_x)$, then

$$M(D,\Omega,\mu\otimes\nu) \subseteq M_u(D,\Lambda,\Delta,\mu,\nu).$$

12.19 LEMMA. The space $M_u(D,\Lambda,\Delta,\mu,\nu)$ is an $M(D,\mu\otimes\nu)$-module and a vector space. Similarly for $M_u^b(D,\Lambda,\Delta,\mu,\nu)$ and $M^b(D,\mu\otimes\nu)$.

Finally, if Λ^* is a continuity structure in ΠE_x^* such that (Λ,Λ^*) is a duality pair, then we have the following commutative diagram of inclusions:

$$
\begin{array}{ccccc}
C(\Omega) & \to & M(\Omega,\mu\otimes\nu) & & \\
\downarrow & & \downarrow & & \\
C_u(\Lambda,\Delta) & \to & M_u(\Lambda,\Delta,\mu,\nu) & & \\
\downarrow & & \downarrow & & \\
C_s(\Lambda,\Delta) & \to & M_{us}(\Lambda,\Delta,\mu,\nu) & \to & M_s(\Lambda,\Delta,\mu,\nu) \\
\downarrow & & \downarrow & & \\
C_w(\Lambda^*,\Delta) & \to & M_{uw}(\Lambda^*,\Delta,\mu,\nu) & \to & M_w(\Lambda^*,\Delta,\mu,\nu)
\end{array}
$$

12.20 REMARK. In deciding how to define compatibility between Ω and Λ,Δ, there are of course possibilities other than that given in 12.5. For example, we could require instead that Ω be contained in any one of the spaces occurring "below" $C_u(\Lambda,\Delta)$ in the previous diagram. In particular, the most important choice for us is $\Omega \subseteq C_s(\Lambda,\Delta)$.

12.21 QUESTIONS. (1) Are all these inclusions proper in general? (2) If $K \in C_u(\Lambda,\Delta)$, is K locally bounded? (3) If $K \in C_u^b(\Lambda,\Delta)$ is $\|K\|$ continuous? In this event, $C_u^b(\Lambda,\Delta)$ is a pre-continuity structure. (4) In general, does there exist Ω such that $C(\Omega) = C_u^b(\Lambda,\Delta)$ as in 12.10, 12.11? If not, is $C_u^b(\Lambda,\Delta)$ the union of the $C(\Omega)$?

13. FINITE-RANK KERNELS

Having discussed various types of continuity and measurability for kernel fields, we next turn our attention to certain special types of kernel fields which will play a fundamental role in Chapter IV. In this section we discuss the vector field analogues of kernel functions whose values are finite-rank operators.

Suppose that $\Lambda, \Delta, \Delta^*$ are continuity structures in ΠE_x, ΠF_y, ΠF_y^* with (Δ, Δ^*) a duality pair. As before, let μ, ν be measures on X, Y.

13.1 DEFINITION. Let f^* and g be elements of ΠF_y^* and ΠE_x respectively. Define the kernel $K_{f^*,g}$ by

$$K_{f^*,g}(x,y)v = \langle v, f^*(y)\rangle g(x) \quad ,$$

for $(x,y) \in X \times Y$ and $v \in F_y$. Then $K_{f^*,g}(x,y)$ is an operator of rank 1 from F_y into E_x which is usually denoted by $f^*(y) \otimes g(x)$. Consider the linear span of the kernels $K_{f^*,g}$ in $\Pi \mathrm{Hom}(F_y, E_x)$. We will call the elements of this subspace the <u>finite-rank kernels</u>. In particular, if K is in the span of the set

$$\{K_{f^*,g} : f^* \in C(\Lambda^*),\ g \in C(\Lambda)\} \quad ,$$

then K will be of special interest to us. This is true also of the finite-rank kernels in the span of the set

$$\{K_{f^*,g} : f^* \in M(\Delta^*, \nu),\ g \in M(\Lambda, \mu)\}.$$

For kernels of these two types, we have the following:

13.2 PROPOSITION. (i) If K is of the first type, then it is strongly continuous, i.e. $K \in \mathcal{C}_s(\Lambda, \Delta)$. (ii) If K is of the second type, then it is strongly measurable, i.e. $K \in \mathcal{M}_s(\Lambda, \Delta, \mu, \nu)$.

Proof. (i) It suffices to assume $K = K_{f^*,g}$, where $f^* \varepsilon C(\Delta^*)$ and $g \varepsilon C(\Lambda)$. If h is an element of Δ, then

$$h_K = \langle h\rho : f^*\rho \rangle g\lambda ,$$

where $g\lambda \varepsilon C(\Lambda\lambda)$ (4.4). Furthermore, $\langle h : f^* \rangle \varepsilon C(Y)$ (8.5), so that

$$\langle h\rho : f^*\rho \rangle = \langle h : f^* \rangle \cdot \rho$$

belongs to $C(X \times Y)$. Thus, h_K is in $C(\Lambda\lambda)$ by 1.9 and K is strongly continuous. Part (ii) is proved similarly.

14. HILBERT-SCHMIDT KERNELS

Finally in this chapter, we discuss the vector field analogues of kernel functions whose values are Hilbert-Schmidt operators. To do this, we must restrict our attention to the Hilbert version of the setting in which we are working. Accordingly, in view of the results of section 6, suppose each E_x, F_y is a Hilbert space, for x in X, y in Y. The vector fields in $\Pi\mathrm{Hom}_2(F_y, E_x)$ will be called the <u>Hilbert-Schmidt kernels</u>. In particular, the finite-rank kernels (in this case) may be viewed as Hilbert-Schmidt kernels.

Given continuity structures Λ, Δ in $\Pi E_x, \Pi F_y$, let Λ^*, Δ^* be as in 8.4, so that (Λ, Λ^*) and (Δ, Δ^*) are duality pairs. In Duflo [1], Duflo considered a certain continuity structure in $\Pi\mathrm{Hom}_2(E_y, E_x)$ (for $X=Y, \Lambda=\Delta$) which we now define (modulo equivalence) in the setting at hand. Let $\Sigma(\Lambda, \Delta)$ denote the linear span in $\Pi\mathrm{Hom}_2(F_y, E_x)$ of the set

$$\{K_{f,g} : f \varepsilon C(\Delta), g \varepsilon C(\Lambda)\} \qquad (f^*=f).$$

Also, let $\Sigma_c(\Lambda, \Delta)$ denote the linear span in $\Sigma(\Lambda, \Delta)$ of the set

$$\{K_{f,g} : f \varepsilon C_c(\Delta), g \varepsilon C_c(\Lambda)\}.$$

(The set $\Sigma_c(\Lambda, \Lambda)$ is the one defined in Duflo [1].)

14.1 PROPOSITION. The spaces $\Sigma(\Lambda, \Delta)$ and $\Sigma_c(\Lambda, \Delta)$ are equivalent continuity structures in $\Pi\mathrm{Hom}_2(F_y, E_x)$.

Proof. We need only verify (ii) and (iii) of 1.1 for $\Sigma(\Lambda, \Delta)$ (recall 1.9). For (ii), observe that if

$$K = \sum_{i=1}^{n} a_i K_{f_i, g_i},$$

where $f_i \ \varepsilon \ C(\Delta)$ and $g_i \ \varepsilon \ C(\Lambda)$, $i=1, \ldots, n$, then

$$\|K(x,y)\|_s^2 = < \sum_{i=1}^{n} a_i K_{f_i, g_i}(x,y), \ \sum_{j=1}^{n} a_j K_{f_j, g_j}(x,y)>$$

$$= \sum_{i,j=1}^{n} a_i \bar{a}_j \langle f_i(y), f_j(y)\rangle \langle g_i(x), g_j(x)\rangle ,$$

where the functions $\langle f_i : f_j\rangle$ and $\langle g_i : g_j\rangle$ are continuous, for $i,j=1,\ldots,n$ (8.5). Thus, $\|K\|_S$ is continuous.

For (iii), we have:

$$K_{f,g}(x,y) = \langle \cdot, f(y)\rangle g(x) ,$$

so that

$$\{K(x,y) : K \varepsilon \Sigma(\Lambda,\Delta)\}$$

is the linear span in $\mathrm{Hom}_2(F_y,E_x)$ of the rank-one operators $f^*(y) \otimes g(x)$. By 1.16, this span is precisely the space of finite-rank operators in $\mathrm{Hom}_2(F_y,E_x)$, so that it is dense in the Schmidt-norm.

The equivalence of $\Sigma_c(\Lambda,\Delta)$ and $\Sigma(\Lambda,\Delta)$ follows from 1.24 , since $\Sigma_c(\Lambda,\Delta) \subseteq \Sigma(\Lambda,\Delta)$.

14.2 REMARK. If $E_x=E$ and $F_y=F$, for all x in X, y in Y , and Λ_0,Δ_0 are as in 1.2, then $\Sigma(\Lambda_0,\Delta_0)$ is not necessarily the space of constant mappings Σ_0 from $X \times Y$ into $\mathrm{Hom}_2(F,E)$. Actually, $\Sigma(\Lambda_0,\Delta_0)$ is the space of constant mappings from $X \times Y$ into the finite-rank operators from F into E. Thus, $\Sigma(\Lambda_0,\Delta_0) \subseteq \Sigma_0$ and it follows from 1.24 that $\Sigma(\Lambda_0,\Delta_0) \sim \Sigma_0$.

In general, we are interested in continuity structures Σ in $\Pi\mathrm{Hom}_2(F_y,E_x)$ which are equivalent to $\Sigma(\Lambda,\Delta)$, i.e. for which $C(\Sigma) = C(\Sigma(\Lambda,\Delta))$. (There appears to be little reason for considering pre-continuity structures here.) In this regard, the following is clear.

14.3 LEMMA. Suppose Λ' and Δ' are also continuity structures in ΠE_x and ΠF_y respectively. If $\Lambda \sim \Lambda'$ and $\Delta \sim \Delta'$, then $\Sigma(\Lambda,\Delta) = \Sigma(\Lambda',\Delta')$.

14.4 COROLLARY. Let Σ denote the span of the set

$$\{K_{f,g} : f\varepsilon\Delta,\ g\varepsilon\Lambda\}$$

in $\Pi\mathrm{Hom}_2(F_y,E_x)$. Then $\Sigma = \Sigma(\Lambda,\Delta)$.

Proof. Recall that $\Lambda\sim C(\Lambda)$ and $\Delta\sim C(\Delta)$ (1.17).

It is important to know how $\Sigma(\Lambda,\Delta)$-continuity fits into the continuity theories developed in sections 10, 11, and 12.

14.5 PROPOSITION. Suppose Σ is a continuity structure in $\Pi\mathrm{Hom}_2(F_y,E_x)$ which is equivalent to $\Sigma(\Lambda,\Delta)$. Then each Σ-continuous kernel is (Λ,Δ)-strongly continuous. Specifically,

$$C(D,\Sigma) \subseteq \mathcal{C}_s^b(D,\Lambda,\Delta)\ , \qquad D\subseteq X\times Y.$$

Proof. Let $K\varepsilon C(D,\Sigma)$ and $h\varepsilon\Delta$. To show: $h_K\varepsilon C(D,\Lambda\lambda)$. Let (x_0,y_0) be a point in D and N a compact neighborhood of y_0 in Y. Choose $0<a<\infty$ such that

$$\|h(y)\| \leq a\ , \qquad a\varepsilon N\cap\rho(D).$$

Given $\varepsilon>0$, there exists J in $\Sigma(\Lambda,\Delta)$ and a neighborhood U' of (x_0,y_0) in $X\times Y$ such that

$$\|K(x,y) - J(x,y)\| \leq \|K(x,y) - J(x,y)\|_s$$

$$< \varepsilon/2a\ , \qquad (x,y)\varepsilon D\cap U'\ ,$$

viewing J as an element of $\mathcal{C}_s(\Lambda,\Delta)$ (13.2). Hence h_J belongs to $C(\Lambda\lambda)$, which implies there exists k in Λ and a neighborhood U of (x_0,y_0) in $X\times Y$ such that

$$U \subseteq U'\cap(X\times N)$$

and

$$\|J(x,y)h(y) - k(x)\| < \varepsilon/2\ , \qquad (x,y)\varepsilon U.$$

For (x,y) in $U\cap D$, we have $y\varepsilon D$ and

$$\begin{aligned}\|K(x,y)h(y)-k(x)\| &\leq \|K(x,y)h(y)-J(x,y)h(y)\| + \|J(x,y)h(y)-k(x)\| \\ &\leq \|K(x,y)-J(x,y)\| \ \|h(y)\| + \|J(x,y)h(y)-k(x)\| \\ &< \varepsilon.\end{aligned}$$

Therefore, $h_K \varepsilon C(\Lambda\lambda)$ and $K \varepsilon \mathcal{C}_s(\Lambda,\Delta)$. Finally, the function $\|K\|$ is locally bounded since $\|K\|_s$ is continuous.

The following fact will be useful in section 17. It shows how $\Sigma(\Lambda,\Delta)$-measurability fits into the measurability theories developed in sections 10, 11, and 12. Let μ,ν be measures on X,Y.

14.6 COROLLARY. Each $(\Sigma(\Lambda,\Delta),\mu\otimes\nu)$-measurable kernel is ultra-strongly (Λ,Δ,μ,ν)-measurable. Specifically, if D is a locally compact subspace of $X\times Y$, then

$$M(D,\Sigma(\Lambda,\Delta),\mu\otimes\nu) \subseteq \mathcal{M}_{us}(D,\Lambda,\Delta,\mu,\nu).$$

14.7 LEMMA. (i) If Λ,Δ are separable, then $\Sigma(\Lambda,\Delta)$ is countably dense. In particular, this is so if Λ,Δ are countably dense. (ii) If Λ,Δ are locally separable, then so is $\Sigma(\Lambda,\Delta)$.

Proof. Let G,H be countable subsets of $C(\Lambda)$, $C(\Delta)$ such that

$$\{g(x) : g\varepsilon G\} , \{f(y) : f\varepsilon H\}$$

are dense in E_x,F_y, for x,y in X,Y. Then

$$\{K_{f,g} : f\varepsilon H , g\varepsilon G\}$$

is a countable subset of $\Sigma(\Lambda,\Delta)$ such that

$$\{K_{f,g}(x,y) : f\varepsilon H , g\varepsilon G\}$$

has dense span in $\mathrm{Hom}_2(F_y,E_x)$, $(x,y) \varepsilon X\times Y$.

We have the following converse of 14.6 (Duflo [1,§3.2]):

14.8 PROPOSITION. If K is weakly $(\Lambda^*,\Delta,\mu,\nu)$-measurable and Λ,Δ are separable, then K is $(\Sigma(\Lambda,\Delta),\ \mu\otimes\nu)$-measurable (for K in $\Pi \mathrm{Hom}_2(F_y,E_x)$).

Proof. $\Sigma(\Lambda,\Delta)$ is countably dense by the set

$$\{K_{f,g} : f\varepsilon H,\ g\varepsilon G\}$$

in the proof of 14.7. Thus, K will be $(\Sigma(\Lambda,\Delta),\ \mu\otimes\nu)$-measurable if the function

$$(x,y) \to \langle K(x,y),\ K_{f,g}(x,y)\rangle$$

on $X\times Y$ is $\mu\otimes\nu$-measurable, for each $f\varepsilon H$, $g\varepsilon G$ (6.20). Since $K_{f,g}(x,y) = f^*(y)\otimes g(x)$, we have:

$$\begin{aligned}\langle K(x,y),K_{f,g}(x,y)\rangle &= \langle K(x,y),f^*(y)\otimes g(x)\rangle\\ &= \langle K(x,y)f(y),g(x)\rangle\ ,\end{aligned}$$

for $f\varepsilon H$, $g\varepsilon G$. By hypothesis, these functions are $\mu\otimes\nu$-measurable (11.13).

14.9 COROLLARY. If Λ,Δ are separable, then the following spaces are all equal to each other in the Hilbert space setting: $M_w(\Lambda^*,\Delta,\mu,\nu)$, $M_{uw}(\Lambda^*,\Delta,\mu,\nu)$, $M_s(\Lambda,\Delta,\mu,\nu)$, $M_{us}(\Lambda,\Delta,\mu,\nu)$ and $M(\Sigma(\Lambda,\Delta),\mu\otimes\nu)$, where the first four spaces are understood to be intersected with $\Pi\mathrm{Hom}_2(F_y,E_x)$.

IV. INTEGRAL OPERATORS AND COMPACTNESS

We are now prepared to define and study the concept of integral operator. In particular, we will be interested in when such an operator is compact, Hilbert-Schmidt, or trace-class. For the purposes of this chapter, assume: $\{E_x\}$, $\{F_y\}$ are fields of Banach spaces over X,Y ; Λ,Δ are continuity structures in ΠE_x, ΠF_y ; μ,ν are measures on X,Y and either X or Y is not a singleton. In addition, wherever required, also assume that Λ^*,Δ^* are continuity structures in ΠE_x^*, ΠF_y^* for which (Λ,Λ^*) , (Δ,Δ^*) are duality pairs.

15. INTEGRAL OPERATORS

In this section we define what it will mean to say that " T is strongly (resp. weakly) an integral operator with kernel K " and study the connection between T and K.

15.1 DEFINITION. Suppose p and q are such that $1 \leqq p,q<\infty$ and T is a bounded operator from $L^q(\Delta,\nu)$ into $L^p(\Lambda,\mu)$.

(i) T is <u>strongly an integral operator</u> if there exists a kernel K having the following properties: For each f in $C_c(\Delta)$, the field f_K is $(\Lambda\lambda,\mu\otimes\nu)$-measurable and for μ-almost all x in X , the mapping

$$f_K(x,\cdot) : Y\to E_x$$

is ν-integrable and

$$Tf(x) = \int K(x,y)f(y)dy.$$

(ii) T is <u>Λ^*-weakly an integral operator</u> if there exists a kernel K having the following properties: For each f in $C_c(\Delta)$ and g^* in $C_c(\Lambda^*)$, the scalar function $\langle f_K : g^*\lambda\rangle$ is $\mu\otimes\nu$-integrable and

$$\langle Tf,g^*\rangle = \int \langle f_K:g^*\lambda\rangle d\mu\otimes\nu .$$

In the event of (i) (resp. (ii)) above, we will say also that K is strongly (resp. Λ^* weakly) a kernel for the integral operator T. Clearly, (i) implies (ii) (Fubini Theorem). We will omit the reference to Λ^* where there is no possibility of confusion.

15.2 LEMMA. (i) If K is strongly a kernel for T, then K is strongly measurable, i.e. $K \epsilon M_s(\Lambda,\Delta,\mu,\nu)$. (ii) If K is weakly a kernel for T, then K is weakly measurable, i.e. $K \epsilon M_w(\Lambda^*,\Delta,\mu,\nu)$.

Proof. These follow essentially from the fact that an element of Δ (resp. Λ^*) is equal to an element of $C_c(\Delta)$ (resp. $C_c(\Lambda^*)$) on a given compact set.

15.3 LEMMA. (i) If K and K' are kernels (strongly or weakly) for T and T' and c is a complex number, then $cK+K'$ is a kernel (strongly or weakly respectively) for $cT+T'$.

(ii) If K is strongly a kernel for T and T', then $T=T'$.

(iii) If K is weakly a kernel for T and T' and $C_c(\Lambda^*)$ is a norming subspace of $L^p(\Lambda,\mu)^*$, then $T=T'$.

Proof. Parts (i) and (ii) are clear. For (iii), observe that for each f in $C_c(\Delta)$, we have by hypothesis:

$$\|Tf - T'f\|_p = \sup_{\substack{g^* \in C_c(\Lambda^*) \\ \|g^*\|_{p^*} \le 1}} |\langle Tf,g^*\rangle - \langle T'f,g^*\rangle|$$

$$= 0 ,$$

Where p^* is such that $1/p + 1/p^* = 1$. Thus, $T=T'$.

15.4 PROPOSITION. If K is weakly a kernel for the zero operator and K is weakly continuous, then K is identically zero on the support of $\mu\otimes\nu$.

Proof. Let (x_0,y_0) be an element of $\mathrm{supp}(\mu\otimes\nu)$ for which $K(x_0,y_0) \neq 0$. Then there exists v in F_{y_0} and v^* in $E^*_{x_0}$ such that:

$$<K(x_0,y_0)v,v^*> > 0.$$

Choose h in Δ and k^* in Λ^* such that $\|h(y_0)-v\|$ and $\|k^*(x_0)-v^*\|$ are small enough to insure that

$$<K(x_0,y_0)h(y_0),\ g^*(x_0)> > 0.$$

By hypothesis, the function $<h_K : k^*\lambda>$ is continuous on $X\times Y$. Thus, there exists $\varepsilon>0$ and a basic open neighborhood $U\times V$ of (x_0,y_0) in $X\times Y$ such that

$$\mathrm{Re}\ <K(x,y)h(y),\ k^*(x)> > \varepsilon\ ,$$

for $(x,y) \in U\times V$. Let $B\times C$ be a product compact neighborhood of (x_0,y_0) which is contained in $U\times V$. Choose ϕ and ψ in $C_c(X)$ and $C_c(Y)$ respectively such that $\mathrm{supp}(\phi\otimes\psi) \subseteq U\times V$, $\phi\otimes\psi=1$ on $B\times C$, and $0\leq\phi\otimes\psi\leq 1$. Then $\phi k^* \in C_c(\Lambda^*)$ and $\psi h \in C_c(\Delta)$, so that by hypothesis:

$$0 = <T(\psi h),\phi k^*> \quad .$$

Consequently,

$$\begin{aligned} 0 &= \mathrm{Re}\ <T(\psi h),\phi k^*> \\ &= \mathrm{Re} \int_{X\times Y} \phi\otimes\psi\ <h_K:k^*\lambda>\ d\mu\otimes\nu \\ &= \int_{U\times V} \phi\otimes\psi\ \mathrm{Re}\ <h_K:k^*\lambda>\ d\mu\otimes\nu \\ &\geq \int_{B\times C} \mathrm{Re}\ <h_K:k^*\lambda>\ d\mu\otimes\nu \\ &< \varepsilon\cdot\mu\otimes\nu\ (B\times C)\ , \end{aligned}$$

which is positive since necessarily $\mu\otimes\nu(B\times C) > 0$ (otherwise, $(x_0,y_0) \notin \mathrm{supp}(\mu\otimes\nu)$). This contradiction proves that $K(x_0,y_0) = 0$.

The following theorem gives a more useful "one-to-one-ness" result than does 15.4.

15.5 THEOREM. Suppose K is weakly a kernel for the integral operator T.

(i) If $T=0$, then $<f_K:g^*\lambda>$ is $\mu\otimes\nu$-null for each f in $C_c(\Delta)$ and g^* in $C_c(\Lambda^*)$.

(ii) If, in addition, Δ and Λ^* are locally separable, then K is locally $\mu\otimes\nu$-null.

Proof. (Due essentially to R.A. Fontenot) (i) First fix f in $C_c(\Delta)$ and g^* in $C_c(\Lambda^*)$ and observe that for each ϕ in $C_c(X)$ and ψ in $C_c(Y)$, we have:

$$0 = \langle T(\psi f), \phi g^* \rangle$$

$$= \int \phi\otimes\psi \langle f_K : g^*\lambda \rangle \, d\mu\otimes\nu .$$

Next, let θ be an element of $C_c(X\times Y)$ and $\varepsilon>0$. Then there exist (Bourbaki [1, p.83]) ϕ_i in $C_c(X)$ and ψ_i in $C_c(Y)$, $i=1,\ldots,n$, such that for $\theta' = \Sigma\phi_i\otimes\psi_i$:

$$\|\theta - \theta'\|_{X\times Y} < \frac{\varepsilon}{\|\langle f_K : g^*\lambda\rangle\|_1} ,$$

when the denominator is not zero. Hence,

$$\begin{aligned} \left|\int\theta\langle f_K:g^*\lambda\rangle \, d\mu\otimes\nu\right| &\le \left|\int(\theta-\theta')\langle f_K:g^*\lambda\rangle \, d\mu\otimes\nu\right| + \left|\int\theta'\langle f_K:g^*\lambda\rangle \, d\mu\otimes\nu\right| \\ &\le \|\theta-\theta'\|_{X\times Y} \|\langle f_K:g^*\lambda\rangle\|_1 + \left|\sum_{i=1}^{n} \int\phi_i\otimes\psi_i \langle f_K:g^*\lambda\rangle \, d\mu\otimes\nu\right| \\ &\le \|\theta-\theta'\|_{X\times Y} \|\langle f_K:g^*\lambda\rangle\|_1 \\ &< \varepsilon , \end{aligned}$$

where ε is arbitrary. Thus,

$$0 = \int\theta\langle f_K:g^*\lambda\rangle \, d\mu\otimes\nu ,$$

for all θ in $C_c(X\times Y)$ as well.

Now define $\alpha = \langle f_K:g^*\lambda\rangle$ and note that α is $\mu\otimes\nu$-integrable and has compact support. Let U be an open subset of $X\times Y$ and assume α is not $\mu\otimes\nu$-null on U. Then it is not difficult to see that there exists $\varepsilon>0$, a $\mu\otimes\nu$-measurable subset S of U (of positive measure) on which either $Re(\alpha) > 2\varepsilon$, $Re(\alpha) < -2\varepsilon$, $Im(\alpha) > 2\varepsilon$ or $Im(\alpha) < -2\varepsilon$. Let β denote either $Re(\alpha)$ or $Im(\alpha)$, depending on which function has the previous property. Then β is a $\mu\otimes\nu$-integrable function on $X\times Y$ satisfying:

$$|\beta(x,y)| > 2\varepsilon$$

and

$$\beta(x,y) < 0 \text{ or } \beta(x,y) > 0, \qquad (x,y) \in S.$$

By the regularity of $\mu\otimes\nu$, there exists a compact subset Q of S for which $\mu\otimes\nu(Q) > 0$ (Hewitt and Stromberg [1, p.178]). Since α is $\mu\otimes\nu$-integrable, we can find $\delta>0$ such that

$$\int_A |\alpha| \, d\mu\otimes\nu < \varepsilon\cdot\mu\otimes\nu(Q),$$

whenever A is a $\mu\otimes\nu$-measurable subset of $X\times Y$ satisfying $\mu\otimes\nu(A)<\delta$ (Hewitt and Stromberg [1, p.176]). Again by the regularity of $\mu\otimes\nu$, there exists an open subset V of $X\times Y$ such that

$$Q \subseteq V \subseteq U$$

and

$$\mu\otimes\nu(V-Q) < \delta.$$

Choose θ in $C_c(X\times Y)$ such that $0\le\theta\le 1$, $supp(\theta)\subseteq V$ and $\theta=1$ on Q. Then by our first observation:

$$0 = \int\theta\langle f_K:g*\lambda\rangle \, d\mu\otimes\nu,$$

which implies that

$$0 = Re \int\theta \langle f_K:g*\lambda\rangle \, d\mu\otimes\nu$$
$$= Im \int\theta \langle f_K:g*\lambda\rangle \, d\mu\otimes\nu$$

i.e.

$$0 = \int\theta \, Re \langle f_K:g*\lambda\rangle \, d\mu\otimes\nu$$
$$= \int\theta \, Im \langle f_K:g*\lambda\rangle \, d\mu\otimes\nu.$$

Consequently,

$$0 = \int_{X\times Y} \theta\beta d\mu\otimes\nu ,$$

so that

$$\begin{aligned} 0 &= \left|\int_V \theta\beta d\mu\otimes\nu\right| \\ &= \left|\int_{V-Q} \theta\beta d\mu\otimes\nu + \int_Q \theta\beta d\mu\otimes\nu\right| \\ &\geq \left| \ \left|\int_Q \theta\beta d\mu\otimes\nu\right| - \left|\int_{V-Q} \theta\beta d\mu\otimes\nu\right| \ \right| \\ &= \left| \ \left|\int_Q \beta d\mu\otimes\nu\right| - \left|\int_{V-Q} \theta\beta d\mu\otimes\nu\right| \ \right| \\ &= \left| \ \int_Q |\beta| d\mu\otimes\nu - \left|\int_{V-Q} \theta\beta d\mu\otimes\nu\right| \ \right|. \end{aligned}$$

However,

$$\begin{aligned} \left| \int_{V-Q} \theta\beta d\mu\otimes\nu\right| &\leq \int_{V-Q} |\beta| d\mu\otimes\nu \\ &\leq \int_{V-Q} |\alpha| d\mu\otimes\nu \\ &< \varepsilon\cdot\mu\otimes\nu(Q) , \end{aligned}$$

since $\mu\otimes\nu(V-Q)<\delta$. On the other hand,

$$\int_Q |\beta| d\mu\otimes\nu > 2\varepsilon\cdot\mu\otimes\nu(Q) ,$$

since $|\beta| > 2\varepsilon$ on Q. Therefore,

$$\begin{aligned} 0 &\geq \int_Q |\beta| d\mu\otimes\nu - \left| \int_{V-Q} \theta\beta d\mu\otimes\nu \right| \\ &> \varepsilon\cdot \mu\otimes\nu(Q) , \end{aligned}$$

which implies that $\mu\otimes\nu(Q) < 0$, i.e. a contradiction. Hence $\langle f_K : g^*\lambda\rangle$ is $\mu\otimes\nu$-null on U which is arbitrary, so that this function is $\mu\otimes\nu$-null. This completes the proof of (i).

(ii) Now suppose also that Δ and Λ^* are locally separable. If A and B are compact subsets of X and Y respectively, then there exists a sequence $\{f_n\}$ in $C(\Delta)$ and a sequence $\{g_m^*\}$ in $C(\Lambda^*)$ such that $\{f_n(y)\}$ is dense in F_y , $y\varepsilon B$, and $\{g_m^*(x)\}$ is dense in E_x^* , $x\varepsilon A$. By a familiar argument, we may assume each f_n and g_m^* has compact support. Define:

$$\alpha_{m,n} = \langle (f_n)_K : g_m^*\lambda\rangle , \qquad m,n=1,2\ldots$$

Then each $\alpha_{m,n}$ is a $\mu\otimes\nu$-null function and there exists a $\mu\otimes\nu$-null subset $N_{m,n}$ of $A\times B$ such that

$$\alpha_{m,n}(x,y) = 0 , \qquad (x,y)\varepsilon A\times B-N_{m,n}.$$

Let

$$N = \bigcup_{m,n=1}^{\infty} N_{m,n} ,$$

so that N is a $\mu\otimes\nu$-null subset of $A\times B$. If (x,y) is an element of $A\times B-N$, then

$$\alpha_{m,n}(x,y) = 0 ,$$

i.e.

$$<K(x,y)f_n(y),g_m^*(x)> = 0 , \qquad m,n=1,2\ldots .$$

This implies that $K(x,y) = 0$. Therefore, K is $\mu\otimes\nu$-null on $A\times B$ and the proof of (ii) is also complete.

For the purpose of the next section, we consider integral operators with kernels which are "(p,q^*)-integrable," where p,q and q^* are such that $1<p$, $q<\infty$ and $1/q+1/q^*=1$. Let $\sigma:Y\times X \to X\times Y$ denote the switching mapping, i.e. $\sigma(y,x) = (x,y)$, for $(y,x)\ \varepsilon\ Y\times X$.

15.6 THEOREM. (i) Suppose first that K is strongly a kernel for T and $\|K\sigma\|_{q^*,p} < \infty$ (recall §7). Suppose also that $\Lambda\lambda$ is ν-measurably restrictive for μ. If $f\ \varepsilon\ L^q(\Lambda,\nu)$, then for μ-almost all x in X, the mapping

$$f_K(x,\cdot) : Y\to E_x$$

is ν-integrable and

$$Tf(x) = \int K(x,y)f(y)dy.$$

Also,

$$\|T\| \le \|K\sigma\|_{q^*,p}.$$

(ii) Now suppose K is weakly a kernel for T and

$$\min(\|K\|_{p,q^*}, \|K\sigma\|_{q^*,p}) < \infty.$$

If $f \in L^q(\Delta,\nu)$ and $g^* \in L^{p^*}(\Lambda^*,\mu)$, then the function $\langle f_K : g^*\lambda\rangle$ is $\mu\otimes\nu$-integrable and

$$\langle Tf, g^*\rangle = \int \langle f_K : g^*\lambda\rangle \, d\mu\otimes\nu.$$

Also, if $C_c(\Lambda^*)$ is a norming subspace of $L^p(\Lambda,\mu)^*$, then

$$\|T\| \leq \min(\|K\|_{p,q^*}, \|K\sigma\|_{q^*,p}).$$

Proof. (i) Let $f \in L^q(\Delta,\nu)$. Since K is strongly measurable (15.2), it follows that $f_K \in M(\Lambda\lambda, \mu\otimes\nu)$. Hence, by hypothesis, $f_K(x,\cdot) \in M(Y,\nu,E_x)$, for μ-almost all x in X. Also, since $\|K\sigma\|_{q^*,p} < \infty$, we have that

$$\int \|K(x,y)\|^{q^*} dy < \infty, \qquad \mu\text{-a.e.}$$

Consequently,

$$\int \|f_K(x,y)\| dy \leq \int^- \|K(x,y)\| \; \|f(y)\| dy$$

$$\leq (\int^- \|K(x,y)\|^{q^*} dy)^{1/q} \|f\|_q$$

(Hölder Inequality, Bourbaki [1, p.12])

$$< \infty, \qquad \mu\text{-a.e.}$$

Thus, for μ-almost all x in X, $f_K(x,\cdot)$ is ν-integrable and its integral is an element of E_x.

Next let $\{f_n\}$ be a sequence in $C_c(\Delta)$ such that $\|f-f_n\|_q \to 0$. Since $\|Tf-Tf_n\|_p \to 0$ also, we may assume (passing to a subsequence if necessary) that $Tf_n(x) \to Tf(x)$, for μ-almost all x in X. For each n, we have:

$$Tf_n(x) = \int K(x,y) f_n(y) dy, \qquad \mu\text{-a.e.}$$

Unioning the appropriate countable collection of μ-null sets, we then have that for x outside a μ-null subset of X:

$$Tf_n(x) = \int K(x,y)f_n(y)dy , \qquad \text{all } n.$$

Therefore, for μ-almost all x in X , it follows that:

$$\begin{aligned}\|Tf(x)-\int K(x,y)f(y)dy\| &\leq \|Tf(x)-Tf_n(x)\| + \|Tf_n(x)-\int K(x,y)f_n(y)dy\| \\ &\quad + \|\int K(x,y)(f_n(y)-f(y))dy\| \\ &\leq \|Tf(x)-Tf_n(x)\| + (\overline{\int}\|K(x,y)\|^{q^*}dy)^{1/q} \|f-f_n\|_q ,\end{aligned}$$

which implies that

$$Tf(x) = \int K(x,y)f(y)dy , \qquad \mu\text{-a.e.}$$

Moreover,

$$\begin{aligned}\|T\| &= \sup_{\substack{f\varepsilon C_c(\Delta) \\ \|f\|_q\leq 1}}\|Tf\|_p \\ &= \sup_{\substack{f\varepsilon C_c(\Delta) \\ \|f\|_q\leq 1}} (\int\|Tf(x)\|^p dx)^{1/p} \\ &= \sup_{\substack{f\varepsilon C_c(\Delta) \\ \|f\|_q\leq 1}} (\int\|\int K(x,y)f(y)dy\|^p dx)^{1/p} \\ &\leq \sup_{\substack{f\varepsilon C_c(\Delta) \\ \|f\|_q\leq 1}} (\overline{\int}(\overline{\int}\|K(x,y)\| \ \|f(y)\|dy)^p dx)^{1/p}\end{aligned}$$

$$\leq \sup_{\substack{f\in C_c(\Delta)\\ \|f\|_q\leq 1}} \left(\int^- \left(\int^- \|K(x,y)\|^{q^*} dy\right)^{p/q^*} \|f\|_q^p dx\right)^{1/p}$$

$$= \sup_{\substack{f\in C_c(\Delta)\\ \|f\|_q\leq 1}} \|f\|_q \left(\int^- \left(\int^- \|K(x,y)\|^{q^*} dy\right)^{p/q^*} dx\right)^{1/p}$$

$$\leq \|K\sigma\|_{q^*,p}.$$

This completes the proof of part (i) of the theorem.

For part (ii), let $f \in L^q(\Delta,\nu)$ and $g^* \in L^{p^*}(\Lambda^*,\mu)$. Since K is weakly measurable (15.2), it follows that the function $\langle f_K : g^*\lambda\rangle$ is $\mu\otimes\nu$-measurable (6.2 and 11.13). Furthermore,

$$\|\langle f_K : g^*\lambda\rangle\|_1 = \int |\langle K(x,y)f(y), g^*(x)\rangle|\, d(x,y)$$

$$= \int\int |\langle K(x,y)f(y), g^*(x)\rangle|\, dxdy$$

(Fubini Theorem, Hewitt and Stromberg [1, p. 386])

$$\leq \int^-\int^- \|K(x,y)\|\ \|f(y)\|\ \|g^*(x)\|\, dxdy$$

$$= \int^- \|f(y)\| \int^- \|K(x,y)\|\ \|g^*(x)\|\, dxdy$$

$$\leq \int^- \|f(y)\| \left(\int^- \|K(x,y)\|^p dx\right)^{1/p} \|g^*\|_{p^*}\, dy$$

(Hölder Inequality, Bourbaki [1, p. 12])

$$= \|g^*\|_{p^*} \int^- \|f(y)\| \left(\int^- \|K(x,y)\|^p dx\right)^{1/p} dy$$

$$\leq \|g^*\|_{p^*} \|f\|_q \|K\|_{p,q^*}.$$

Similarly, by iterating the double integral in the opposite order, we obtain:

$$\|\langle f_K : g^*\lambda\rangle\|_1 \leq \|g^*\|_{p^*} \|f\|_q \|K\sigma\|_{q^*,p}.$$

Thus, the function $<f_K:g^*\lambda>$ is $\mu\otimes\nu$-integrable. Now suppose $\{g_n^*\}$ is a sequence in $C_c(\Lambda^*)$ and $\{f_n\}$ is a sequence in $C_c(\Lambda)$ such that

$$\|f - f_n\|_q \to 0 \text{ and } \|g^* - g_n^*\|_{p^*} \to 0.$$

Then:

$$<Tf,g^*> = \lim_{n\to\infty} <Tf_n,g_n^*>$$

$$= \lim_{n\to\infty} \int<(f_n)_K:g_n^*\lambda> d\mu\otimes\nu.$$

On the other hand,

$$\begin{aligned}|\int(<f_K:g^*\lambda>-<(f_n)_K:g_n^*\lambda>)d\mu\otimes\nu| &\leq |\int(<f_K:g^*\lambda>-<f_K:g_n^*\lambda>)d\mu\otimes\nu| \\ &\quad + |\int(<f_K:g_n^*\lambda>-<(f_n)_K:g_n^*\lambda>)d\mu\otimes\nu| \\ &\leq \|<f_K:(g^*-g_n^*)\lambda>\|_1 + \|<(f-f_n)_K:g_n^*\lambda>\|_1 \\ &\leq \min(\|K\sigma\|_{q^*,p},\|K\|_{p,q^*})\cdot \\ &\quad [\|f\|_q \|g^*-g_n^*\|_{p^*} + \|f-f_n\|_q \|g_n^*\|_{p^*}] ,\end{aligned}$$

which converges to zero since the sequence $\{g_n^*\}$ is bounded. Therefore,

$$\lim_{n\to\infty} \int <(f_n)_K:g_n^*\lambda> d\mu\otimes\nu = \int<f_K:g^*\lambda> d\mu\otimes\nu$$

and the first part of (ii) is proved.

Next suppose $C_c(\Lambda^*)$ is a norming subspace of $L^p(\Lambda,\mu)^*$. Then

$$\|T\| = \sup_{\substack{f\varepsilon C_c(\Delta)\\ \|f\|_q\leq 1}} \sup_{\substack{g^*\varepsilon C_c(\Lambda^*)\\ \|g^*\|_{p^*}\leq 1}} |<Tf,g^*>|$$

$$= \sup_{\substack{f\varepsilon C_c(\Delta)\\ \|f\|_q\leq 1}} \sup_{\substack{g^*\varepsilon C_c(\Lambda^*)\\ \|g^*\|_{p^*}\leq 1}} |\int<f_K:g^*\lambda>d\mu\otimes\nu|$$

$$\leq \sup_{\substack{f\in C_c(\Delta)\\ \|f\|_q\leq 1}} \sup_{\substack{g^*\in C_c(\Lambda^*)\\ \|g^*\|_{p^*}\leq 1}} \|\langle f_K : g^*\lambda\rangle\|_1$$

$$\leq \sup_{\substack{f\in C_c(\Delta)\\ \|f\|_q\leq 1}} \sup_{\substack{g^*\in C_c(\Lambda^*)\\ \|g^*\|_{p^*}\leq 1}} \min(\|K\|_{p,q^*},\|K\sigma\|_{q^*,p})\|f\|_q \|g^*\|_{p^*}$$

(by previous computation), so that

$$\|T\| \leq \min (\|K\|_{p,q^*},\|K\sigma\|_{q^*,p}).$$

15.7 EXAMPLE. Let $f^*\varepsilon M(\Delta^*,\nu)$ and $g\varepsilon M(\Lambda,\mu)$ and define K to be the kernel $K_{f^*,g}$ in section 13. For $f\varepsilon M(\Delta,\nu)$ we have:

$$K(x,y)f(y) = \langle f(y),f^*(y)\rangle\ g(x) ,$$

for $(x,y) \varepsilon X\times Y$. Thus, the vector field f_K is $(\Lambda\lambda,\mu\otimes\nu)$-measurable. Suppose in particular that $g\varepsilon L^p(\Lambda,\mu)$ and $f^*\varepsilon L^{q^*}(\Delta^*,\nu)$ and define $T=T_{f^*,g}$ by:

$$T : L^q(\Delta,\nu) \to L^p(\Lambda,\mu) ,$$

where

$$\begin{aligned} T(f) &= \langle f,f^*\rangle g \\ &= (\textstyle\int\langle f:f^*\rangle d\nu)g , \qquad\qquad f\varepsilon L^q(\Delta,\nu). \end{aligned}$$

15.8 PROPOSITION. T is a bounded linear operator from $L^q(\Delta,\nu)$ into $L^p(\Lambda,\mu)$ with

$$\|T\| \leq \|f^*\|_{q^*} \|g\|_p.$$

Furthermore, K is strongly a kernel for T.

Proof. The first part is clear, since

$$\|Tf\|_p = |\langle f,f^*\rangle|\ \|g\|_p$$

$$\leq \|f^*\|_{q^*} \|g\|_p \|f\|_q , \qquad f \in L^q(\Delta,\nu).$$

For the second part, observe that for f in $C_c(\Delta)$ the function $<f:f^*>g(x)$ is ν-integrable for μ-almost all x in X. Also, for such x, we have:

$$Tf(x) = (\int <f(y),f^*(y)>dy)g(x)$$

$$= \int <f(y),f^*(y)>g(x)dy$$

$$= \int K_{f^*,g}(x,y)f(y)dy ,$$

which completes the proof.

15.9 EXAMPLE. Suppose Ω is a "strong" (Λ,Δ)-pre-continuity structure (12.20). Let $\theta \in C_c(X\times Y)$ and $J \in \Omega$, so that $K=\theta J \in C_c(\Omega)$. Given f in $C_c(\Delta)$, fix x in X and consider the mapping of Y into E_x given by

$$y \to \theta(x,y)J(x,y)f(y).$$

Since $\Omega \subseteq \mathcal{C}_s(\Lambda,\Delta)$, it follows (10.5) that f_J belongs to $C_c(\Lambda\lambda)$; thus, θf_J also belongs to $C_c(\Lambda\lambda)$. Therefore, the above mapping belongs to $C_c(Y,E_x)$ and its integral

$$\int \theta(x,y)J(x,y)f(y)dy$$

is an element of E_x. In this manner, we obtain a vector field, denoted by Tf, where $Tf(x)$ is the previous integral.

15.10 PROPOSITION. For each f in $C_c(\Delta)$, the field Tf belongs to $C_c(\Lambda)$.

Proof. Let Q_1 and Q_2 be compact subsets of X and Y such that $\text{supp}(\theta) \subseteq Q_1\times Q_2$. Also let A be an arbitrary compact subset of X. On the compact set $A\times Q_2$, the field θf_J is the uniform limit of fields of the form

$$\sum_{j=1}^{n} \tau_j(h_j\lambda) ,$$

where $\tau_j \varepsilon C(X\times Y)$ and $h_j \varepsilon \Lambda$, $j=1,\ldots,n$ (1.11). Letting U_1 and U_2 be pre-compact open neighborhoods of Q_1 and Q_2 in X and Y respectively, we may assume $\tau_j \varepsilon C_c(X\times Y)$ with $\mathrm{supp}(\tau_j) \subseteq U_1\times U_2$, $j=1,\ldots,n$. Hence, each τ_j in turn is the uniform limit on $\overline{U}_1\times\overline{U}_2$ of functions of the form

$$\sum_{i=1}^{n} \phi_i \otimes \psi_i ,$$

where $\phi_i \varepsilon C_c(X)$, $\psi_i \varepsilon C_c(Y)$, $\mathrm{supp}(\phi_i) \subseteq \overline{U}_i$ and $\mathrm{supp}(\psi_i) \subseteq \overline{U}_2$, $i=1,\ldots,n$. Thus, on $A\times Q_2$, ϕf_J is the uniform limit of fields of the form

$$\sum_{j=1}^{n} \phi_i \otimes \psi_j (h_j \lambda) ,$$

where $\phi_j \varepsilon C_c(X)$, $\psi_j \varepsilon C_c(Y)$ and $h_j \varepsilon \Lambda$, $j=1,\ldots,n$. Given $\varepsilon>0$, let g in $\Pi E_{x,y}$ be a field of this form satisfying:

$$\|\theta(x,y)J(x,y)f(y) - g(x,y)\| < \frac{\varepsilon}{a} , \qquad (x,y)\varepsilon A\times Q_2 ,$$

where $a>0$ is chosen so that $\mu(Q_2)\leq a$. Let k denote the field in $C(X)\cdot\Lambda$ obtained from g as follows:

$$k(x) = \sum_{j=1}^{n} a_j \phi_j(x) h_j(x) , \qquad x\varepsilon X ,$$

where

$$a_j = \int_{Q_2} \psi_j(y)dy , \qquad j=1,\ldots,n.$$

Then for x in Δ , we have:

$$\begin{aligned}
\|Tf(x)-k(x)\| &= \|\int_Y \theta(x,y)J(x,y)f(y)dy - \sum_{j=1}^{n} a_j\phi_j(x)h_j(x)\| \\
&= \|\int_{Q_2} \theta(x,y)J(x,y)f(y)dy - \sum_{j=1}^{n} \int_{Q_2} \phi_j(x)\psi_j(y)h_j(x)dy\| \\
&\leq \int_{Q_2} \|\theta(x,y)J(x,y)f(y) - g(x,y)\|dy \\
&< \varepsilon.
\end{aligned}$$

Consequently, since A is arbitrary, Tf is Λ-continuous (1.11). This completes the proof since

$$\text{supp}(Tf) \subseteq \lambda(\text{supp}(\theta)).$$

15.11 PROPOSITION. For each choice of p and q satisfying $1<p,q<\infty$, the mapping T extends to a bounded linear operator from $L^q(\Delta,\nu)$ into $L^p(\Lambda,\mu)$ with

$$\|T\| \leq \|(\theta J)\circ\sigma\|_{q^*,p}.$$

Furthermore, θJ is strongly a kernel for T.

Proof. The norm inequality is proved as in the first part of 15.6. The rest of the proposition follows from 15.9 and 15.10.

In sections 17 and 18 we will want to consider adjoints of integral operators and their kernels in the Hilbert space context. Thus, we direct our attention to this consideration for the remainder of this section. Accordingly, suppose E_x and F_y are Hilbert spaces for each x in X and y in Y (6.13). In this case, $L^2(\Lambda,\mu)$, $L^2(\Delta,\nu)$ are Hilbert spaces and $L^2(\Lambda^*,\mu)$, $L^2(\Delta^*,\nu)$ are equivalent to $L^2(\Lambda,\mu)$, $L^2(\Delta,\nu)$ (8.13). In particular, $C_c(\Lambda^*)$, $C_c(\Delta^*)$ are norming subspaces of $L^2(\Lambda,\mu)^*$, $L^2(\Delta,\nu)^*$. (We are assuming $\Lambda^* = \langle\cdot\,,\Lambda\rangle$ and $\Delta^* = \langle\cdot,\Delta\rangle$.) In this setting, strong integral operators are Λ^*-weak integral operators. Thus, it will be convenient to speak simply of integral operators in the Hilbert space context (here and in §§17,18).

Suppose $T:L^2(\Delta,\nu) \to L^2(\Lambda,\mu)$ is a bounded operator. Recall that the adjoint of T is the unique bounded operator $T^*:L^2(\Lambda,\mu) \to L^2(\Delta,\nu)$ satisfying:

$$\langle T^*g,f\rangle = \langle f,Tg\rangle,$$

for $f\in C_c(\Delta)$ and $g\in C_c(\Lambda)$. Thus, if T^* is to be an integral operator, a kernel for T^* must be a field in $\Pi\text{Hom}(E_x,F_y)$. Accordingly, consider the field $\{\text{Hom}(E_x,F_y)\}$ of Banach spaces over $Y\times X$, where $\text{Hom}(E_x,F_y)$ is the space over (y,x) in $Y\times X$.

Now let K be a kernel in $\Pi\mathrm{Hom}(F_y,E_x)$. Then as above, for each (x,y) in $X\times Y$, $K(x,y)^*$ is the unique bounded operator from E_x into F_y determined by

$$\langle K(x,y)^*v,w\rangle = \langle v,K(x,y)w\rangle , \qquad w\in F_y , \; v\in E_x .$$

Therefore, as usual, we obtain a kernel K^* in $\Pi\mathrm{Hom}(E_x,F_y)$ by defining:

$$K^*(y,x) = K(x,y)^* , \qquad (y,x)\in Y\times X.$$

15.12 LEMMA. If $T:L^2(\Delta,\nu) \to L^2(\Lambda,\mu)$ is an integral operator with kernel K, then $T^*:L^2(\Lambda,\mu) \to L^2(\Delta,\nu)$ is an integral operator with kernel K^*.

Proof. Let $f\in C_c(\Delta), g\in C_c(\Lambda)$. Then

$$\overline{\langle f_K:g^*\lambda\rangle}\circ\sigma = \langle g_{K^*}:f^*\lambda\rangle$$

as functions on $Y\times X$. This function is $\nu\otimes\mu$-integrable and

$$\begin{aligned}
\langle T^*g,f\rangle &= \langle g,Tf\rangle \\
&= \overline{\langle Tf,g\rangle} \\
&= \int\overline{\langle f_K:g^*\lambda\rangle}d\mu\otimes\nu \\
&= \int\overline{\langle f_K:g^*\lambda\rangle}\circ\sigma d\nu\otimes\mu \\
&= \int\langle g_{K^*}:f^*\lambda\rangle d\nu\otimes\mu .
\end{aligned}$$

15.13 LEMMA. Suppose $X=Y$, $E_x=F_x$, $x\in X$, $\Lambda=\Delta$ and $\mu=\nu$. Suppose also that $T:L^2(\Lambda,\mu) \to L^2(\Lambda,\mu)$ is an integral operator with weakly continuous kernel K, i.e. $K\in C_w(\Lambda^*,\Lambda)$. If T is a positive operator, then $K(x,x)$ is a positive operator on E_x, for each x in the support of μ.

Proof. See Duflo [1, Théorème 3.3.1].

15.14 QUESTIONS. (1) Does the weak integral operator property really depend on the choice of Λ^* or, more accurately, on the equivalence class of Λ^*? (2) What are the strong analogues of 15.4 and 15.6?

16. COMPACT OPERATORS

Given an integral operator T with kernel K, we may ask: What conditions on K will insure that T is a compact operator? Our goal in this section is to extend a familiar classical result on compact integral operators to the context of vector fields (see Dunford and Schwartz [1, p.518, Ex.52]).

Suppose that Ω is a strong (Λ,Δ)-pre-continuity structure in $\Pi\mathrm{Hom}(F_y,E_x)$ (12.20). Given $1<p,q<\infty$, recall (15.11) that for each θ in $C_c(X\times Y)$ and J in Ω, there exists an integral operator

$$T:L^q(\Delta,\nu) \to L^p(\Lambda,\mu)$$

with strong kernel θJ. In particular, this is the case for $\theta=\phi\otimes\psi$, where $\phi\in C_c(X), \psi\in C_c(Y)$.

16.1 DEFINITION. The pre-continuity structure Ω is of <u>compact type for (μ,ν)</u> if for each $1<p, q<\infty$, $J\in\Omega$, $\phi\in C_c(X)$ and $\psi\in C_c(Y)$, the operator T corresponding to $(\phi\otimes\psi)J$ is compact.

16.2 REMARK. It is easy to see that the choices of μ and ν do matter. For example suppose $X=Y=\{1,2\}$, $E_1=F_1=\mathbb{C}$ and $E_2=F_2$ is any infinite dimensional Hilbert space. Let $\Lambda=\Delta=E_1\times E_2$. Also, let μ_1 and μ_2 be the Dirac measures at 1 and 2 respectively. Define:

$$B_{x,y} = \delta_{x,y}\mathrm{Hom}(E_y,E_x), \qquad (x,y) \in X\times X,$$

where $\delta_{x,y}$ is the Kronecker delta. If we let $\Omega=\Pi B_{x,y}$, then Ω is of compact type for (μ_1,μ_1) but not (μ_2,μ_2).

16.3 LEMMA. Suppose Ω is of compact type for (μ,ν). Then for each choice of K in $C_c(X\times Y)\cdot\Omega$ and $1<p,q<\infty$, the corresponding integral operator is compact.

Proof. It suffices to consider K of the form θJ, for θ a non-zero element of $C_c(X\times Y)$ and J in Ω. Let A and B be compact subsets of X and Y respectively such that $\text{supp}(\theta) \subseteq A\times B$. Given $\varepsilon>0$, choose θ_ε in $C_c(X)\otimes C_c(Y)$ such that

$$\|\theta - \theta_\varepsilon\|_{X\times Y} < \frac{\varepsilon}{\|\gamma_A\otimes\gamma_B J\circ\sigma\|_{q^*,p}} .$$

If T_ε is the integral operator corresponding to $\theta_\varepsilon J$, then T_ε is compact and

$$\begin{aligned}\|T - T_\varepsilon\| &\leq \|(\theta-\theta_\varepsilon)J\circ\sigma\|_{q^*,p} \\ &\leq \|\theta-\theta_\varepsilon\|_{X\times Y}\,\|\gamma_A\otimes\gamma_B J\circ\sigma\|_{q^*,p} \\ &< \varepsilon,\end{aligned} \tag{15.11}$$

so that T is compact.

For the next theorem, note that $\Omega\sigma = \{J\sigma : J\varepsilon\Omega\}$ is a pre-continuity structure in $\prod_{Y\times X} \text{Hom}(F_y,E_x)$ and that section 7 is valid for pre-continuity structures as well.

16.4 THEOREM. Suppose Ω is of compact type for (μ,ν), $1<p,q<\infty$ and $T:L^q(\Delta,\nu) \to L^p(\Lambda,\mu)$ is a bounded operator.

(i) If T is strongly an integral operator with kernel K and $K\sigma \in L^{q^*,p}(\Omega\sigma,\nu,\mu)$, then T is compact.

(ii) Suppose also that $C_c(\Lambda^*)$ is a norming subspace of $L^p(\Lambda,\mu)^*$. If T is weakly an integral operator with kernel K and $K\varepsilon L^{p,q^*}(\Omega,\mu,\nu)$ or $K\sigma \in L^{q^*,p}(\Omega\sigma,\nu,\mu)$, then T is compact.

Proof. (i) Let $\varepsilon>0$. Since $C_c(Y\times X)\cdot\Omega\sigma = (C_c(X\times Y)\cdot\Omega)\sigma$ is dense in $L^{q^*,p}(\Omega\sigma,\nu,\mu)$ (7.4), there exists K_ε in $C_c(X\times Y)\cdot\Omega$ such that

$$\|K\sigma-K_\varepsilon\sigma\|_{q^*,p} < \varepsilon.$$

Also, K_ε is a kernel for a compact operator

$$T_\varepsilon : L^q(\Delta,\nu) \to L^p(\Lambda,\mu) \qquad (16.3).$$

Thus, $K-K_\varepsilon$ is a kernel for the operator $T-T_\varepsilon$ (15.3) and

$$\|T-T_\varepsilon\| < \varepsilon . \qquad (15.6)$$

Since ε is arbitrary, T is compact. Part (ii) is proved similarly.

16.5 COROLLARY. Suppose $p=q=2$ and $K \varepsilon L^2(\Omega,\mu\otimes\nu)$. Then T is compact.

Proof. In this case, $L^2(\Omega,\mu\otimes\nu) = L^{2,2}(\Omega,\mu,\nu) = L^{2,2}(\Omega\sigma,\nu,\mu)$ by Fubini's Theorem (7.3).

16.6 PROPOSITION. Suppose $E_x=E$, $F_y=F$, for all x in X, y in Y, $1<p,q<\infty$ and $T:L^q(Y,\nu,F) \to L^p(X,\mu,E)$ is a bounded operator.

(i) If T is strongly an integral operator with kernel K and

$$K\sigma \ \varepsilon\ L^{q^*,p}(Y\times X,\ \nu\otimes\mu,\ \mathrm{Hom}_\infty(F,E)) ,$$

then T is compact.

(ii) Suppose also that $C_c(X,E^*)$ is a norming subspace of $L^p(X,\mu,E)^*$.

If T is weakly (usual sense) an integral operator with kernel K and

$$K \ \varepsilon\ L^{p,q^*}(X\times Y,\ \mu\otimes\nu,\ \mathrm{Hom}_\infty(F,E))$$

or

$$K\sigma \ \varepsilon\ L^{q^*,p}(Y\times X,\ \nu\otimes\mu,\ \mathrm{Hom}_\infty(F,E)) ,$$

then T is compact.

Proof. Let Ω denote the space of constant mappings from $X\times Y$ into $\mathrm{Hom}_\infty(F,E)$. By the previous theorem, it suffices to show that Ω is of compact type for (μ,ν). Let ϕ and ψ be non-zero elements of $C_c(X)$ and $C_c(Y)$ respectively, J an element of Ω and R an element of $\mathrm{Hom}_\infty(F,E)$ such that $J(x,y) = R$, $(x,y)\ \varepsilon\ X\times Y$. The kernel $\phi\otimes\psi J$ defines an integral operator T as follows:

$$Tf(x) = \phi(x)\int\psi(y)Rf(y)dy$$

$$= \phi(x)R(\int\psi(y)f(y)dy) ,$$

where the integral is a vector in E. Define:

$$P : E \to L^p(X,\mu,E)$$

by

$$P(v) = \phi\otimes v , \qquad v\varepsilon E ,$$

and

$$S : L^q(Y,\nu,F) \to F$$

by

$$S(f) = \int\psi f d\nu , \qquad f\varepsilon L^q(Y,\nu,F).$$

Then $T=PRS$ which is compact since R is.

16.7 REMARK. Before leaving this section, we should observe that the operators defined in 15.7 and 15.8 are compact since they are of finite-rank.

16.8 QUESTIONS. (1) If K is a kernel for a compact integral operator, must $K(x,y)$ be a compact operator for locally all (x,y) in $X\times Y$?
(2) Is Ω of compact type for (μ,ν) if for each J in Ω , $J(x,y)$ is a compact operator for all (x,y) in $\mathrm{supp}(\mu\otimes\nu)$?

17. HILBERT-SCHMIDT OPERATORS

In this section we assume $L^p(\Lambda,\mu)$ and $L^q(\Delta,\nu)$ are Hilbert spaces (recall section 6) and ask what conditions on the kernel K of an integral operator T will make T Hilbert-Schmidt. In the classical case, it is well-known that all Hilbert-Schmidt operators on $L^2(X,\mu,E)$ are integral operators with kernels belonging to $L^2(X\times X,\mu\otimes\mu,\mathrm{Hom}_2(E))$. In fact, this correspondence determines a unitary transformation of $L^2(X\times X,\mu\otimes\mu,\mathrm{Hom}_2(E))$ onto $\mathrm{Hom}_2(L^2(X,\mu,E))$, i.e. these Hilbert spaces are isomorphic. We shall see that this correspondence extends to the context of vector fields in a canonical fashion, thereby giving a complete description of the Hilbert-Schmidt operators on the Hilbert space $L^2(\Delta,\nu)$. More generally, we consider the Hilbert-Schmidt operators from $L^2(\Delta,\nu)$ into $L^2(\Lambda,\mu)$. The main results of this section are originally due to Duflo for the case $X=Y$, $E_x=F_x$, $x\varepsilon X$, and $\Delta=\Lambda$ (Duflo [1, §3.2]).

As suggested by the relevant results of section 6, suppose E_x,F_y are Hilbert spaces for $x\varepsilon X, y\varepsilon Y$, so that $L^2(\Lambda,\mu)$ and $L^2(\Delta,\nu)$ are Hilbert spaces. Let $\Sigma(\Lambda,\Delta)$ and $\Sigma_c(\Lambda,\Delta)$ be as in section 14. Since Λ and Δ will be fixed, it will be convenient to write Σ and Σ_c respectively for these spaces in what follows. For f in $L^2(\Delta,\nu)$ and g in $L^2(\Lambda,\mu)$ let $K_{f,g}$ amd $T_{f,g}$ be as in 15.7. (Technically, we should write f^* instead of f here, but the meaning is clear.) The following are classical results in the theory of Hilbert spaces.

17.1 LEMMA. The subspace of $\mathrm{Hom}(L^2(\Delta,\nu),L^2(\Lambda,\mu))$ generated by the set

$$\{T_{f,g} : f\varepsilon L^2(\Delta,\nu), g\varepsilon L^2(\Lambda,\mu)\}$$

(finite-rank operators) is dense in the compact operators (operator norm) and in the Hilbert-Schmidt operators (Schmidt-norm) from $L^2(\Delta,\nu)$ into $L^2(\Lambda,\mu)$. Furthermore,

$$\|T_{f,g}\| \leq \|T_{f,g}\|_s = \|f\|_2 \|g\|_2$$

and

$$\langle T_{f,g}, T_{h,k}\rangle = \langle f,h\rangle \langle g,k\rangle ,$$

for $f,h \in L^2(\Delta,\nu)$ and $g,k \in L^2(\Lambda,\mu)$.

Next observe that $\Sigma_c \subseteq C_c(\Sigma)$, i.e. Σ_c is contained in $L^2(\Sigma,\mu\otimes\nu)$.

17.2 LEMMA. For $f,h \in C_c(\Delta)$ and $g,k \in C_c(\Lambda)$, we have:

(i) $$\|K_{f,g}\|_2 = \|f\|_2 \|g\|_2 .$$

(ii) $$\langle K_{f,g}\, K_{h,k}\rangle = \langle f,h\rangle \langle g,k\rangle .$$

17.3 PROPOSITION. The subspace Σ_c of $L^2(\Sigma,\mu\otimes\nu)$ is dense.

Proof. Fix $\varepsilon>0$ and let K be a non-zero element of $C_c(\Sigma)$. Let Q_1 and Q_2 be compact subsets of X and Y respectively such that $Q = Q_1\times Q_2$ contains the support of K. Let $V = V_1\times V_2$, where V_i is a precompact open neighborhood of Q_i in the appropriate space X or Y, $i = 1,2$. Since K is Σ-continuous, there exists K_3 in $C(\Sigma)$ of the form

$$K_3 = \sum_{i=1}^{n} \psi_i J_i , \qquad \psi_i \in C(X\times Y), J_i \in \Sigma,$$

such that

$$\|K(x,y)-K_3(x,y)\|_s < \frac{\varepsilon}{2\sqrt{2}\, \mu\otimes\nu(Q)^{1/2}} ,$$

for (x,y) in V (1.11). Choose open neighborhoods U_j of Q_j such that $U_j \subseteq V_j$, $j=1,2$, and

$$\min\{\mu(U_1-Q_1), \nu(U_2-Q_2)\} < \frac{\varepsilon}{\sqrt{8}} \max_{(x,y)\in V} [\|K(x,y)\|_s + \|K_3(x,y)\|_s] .$$

Then, for $U = U_1\times U_2$, we have:

$$\mu\otimes\nu(U-Q) = \mu(U_1-Q_1)\cdot\nu(U_2-Q_2)$$

$$< \frac{\varepsilon^2}{8} \max_{(x,y)\varepsilon V} [\|K(x,y)\|_s + \|K_3(x,y)\|_s]^2 .$$

Next choose β in $C_c(X\times Y)$ such that $0\leq\beta\leq 1$, $\text{supp}(\beta)\subseteq U$ and $\beta=1$ on Q. Let $\phi_i=\beta\psi_i$, $i=1,\ldots,n$. Then $\phi_i \varepsilon C_c(X\times Y)$, $\text{supp}(\phi_i)\subseteq U$ and $\phi_i=\psi_i$ on Q , $i=1,\ldots,n$. Define:

$$K_2 = \sum_{i=1}^{n} \phi_i J_i ,$$

so that $K_2 \varepsilon C_c(\Sigma)$, $\text{supp}(K_2)\subseteq U$ and $K_2=K_3$ on Q. Let $c>0$ be such that

$$c > \max_{\substack{(x,y)\varepsilon U \\ 1\leq i\leq n}} \|J_i(x,y)\|_s .$$

For each $i=1,\ldots,n$, there exists θ_i in $C_c(X) \otimes C_c(Y)$ such that $\text{supp}(\theta_i) \subseteq U$ and

$$\|\phi_i-\theta_i\|_U < \frac{\varepsilon}{2nc\mu\otimes\nu(U)^{\frac{1}{2}}} , \qquad i=1,\ldots,n.$$

Finally, let

$$K_1 = \sum_{i=1}^{n} \phi_i J_i .$$

Then clearly, $K_1 \varepsilon \Sigma_c$ and

$$\|K-K_1\|_2 \leq \|K-K_2\|_2 + \|K_2-K_1\|_2 ,$$

where

$$\|K-K_2\|_2^2 = \int_{X\times Y} \|K-K_2\|_s^2 \, d\mu\otimes\nu$$

$$= \int_U \|K-K_2\|_s^2 \, d\mu\otimes\nu$$

$$= \int_{U-Q} \|K-K_2\|_s^2 \, d\mu\otimes\nu + \int_Q \|K-K_2\|_s^2 \, d\mu\otimes\nu$$

$$\leq \int_{U-Q} [\|K\|_s + \|K_3\|_s]^2 \, d\mu\otimes\nu + \int_Q \|K-K_2\|_s^2 \, d\mu\otimes\nu$$

(since $\|K_2\|_s \leq \|K_3\|_s$)

$$\leq [\|K\|_s+\|K_3\|_s]_V^2 \, \mu\otimes\nu(U-Q) + \int_Q \|K-K_2\|_s^2 \, d\mu\otimes\nu$$

$$< \varepsilon\, 2/8 + \varepsilon\, 2/8$$

$$= \varepsilon\, 2/4$$

and

$$\begin{aligned}
\|K_2-K_1\|_2^2 &= \int_{X\times Y}\|K_2-K_1\|_s^2 \, d\mu\otimes\nu \\
&= \int_U \|K_2-K_1\|_s^2 \, d\mu\otimes\nu \\
&= \int_U \|\sum_{i=1}^n (\phi_i-\theta_i)J_i\|_s^2 \, d\mu\otimes\nu \\
&\le \int_U [\sum_{i=1}^n |\phi_i-\theta_i| \, \|J_i\|_s]^2 \, d\mu\otimes\nu \\
&\le \int_U [\sum_{i=1}^n \frac{\varepsilon}{2nc\mu\otimes\nu(U)^{\frac{1}{2}}} \|J_i\|_s]^2 \, d\mu\otimes\nu \\
&\le \int_U [\sum_{i=1}^n \frac{\varepsilon}{2n\mu\otimes\nu(U)^{\frac{1}{2}}}]^2 \, d\mu\otimes\nu \\
&= \frac{\varepsilon^2}{4} .
\end{aligned}$$

Therefore, $\|K-K_1\|_2 < \varepsilon$ and the proof is complete.

The next step is to construct a unitary transformation

$$S:L^2(\Sigma,\mu\otimes\nu) \to \mathrm{Hom}_2(L^2(\Delta,\nu),L^2(\Lambda,\mu)) .$$

For $f \in C_c(\Delta)$ and $g \in C_c(\Lambda)$, we have $K_{f,g} \in \Sigma_c$ and we define $S(K_{f,g})$ to be the operator $T_{f,g}$ in $\mathrm{Hom}_2(L^2(\Delta,\nu), L^2(\Lambda,\mu))$. Now suppose $K \in \Sigma_c$ and

$$K = \sum_{i=1}^n K_{f_i,g_i} = \sum_{j=1}^m K_{h_j,k_j} ,$$

where $h_j \in C_c(\Delta)$, $k_j \in C_c(\Lambda)$, $j=1,\ldots,m$. For $f \in C_c(\Delta)$ and $g \in C_c(\Lambda)$, it is a routine matter to show that:

$$\begin{aligned}
\langle(\sum_{i=1}^n T_{f_i,g_i})f,g\rangle &= \sum_{i=1}^n \langle f,f_i\rangle \langle g_i,g\rangle \\
&= \int\langle f_K : g\lambda\rangle \, d\mu\otimes\nu
\end{aligned}$$

$$= \sum_{j=1}^{m} \langle f,h_j\rangle \langle k_j,g\rangle$$

$$= \langle (\sum_{j=1}^{m} T_{h_j,k_j})f,g\rangle .$$

Consequently, S extends by linearity to a well-defined mapping of Σ_c into $\mathrm{Hom}_2(L^2(\Delta,\nu),L^2(\Lambda,\mu))$. Furthermore, if $K \in \Sigma_c$, then:

$$\|S(K)\|_s^2 = \|\sum_{i=1}^{n} T_{f_i,g_i}\|_s^2$$

$$= \sum_{i=1}^{n} \sum_{j=1}^{m} \langle T_{f_i,g_i}, T_{f_j,g_j}\rangle$$

$$= \sum_{i=1}^{n} \sum_{j=1}^{m} \langle f_i,f_j\rangle \langle g_i,g_j\rangle$$

$$= \sum_{i=1}^{n} \sum_{j=1}^{m} \langle K_{f_i,g_i}, K_{f_j,g_j}\rangle$$

$$= \|K\|_2^2 ,$$

so that S is isometric. Hence, S can be extended again to an isometry of $L^2(\Sigma,\mu\otimes\nu)$ which is onto $\mathrm{Hom}_2(L^2(\Delta,\nu), L^2(\Lambda,\mu))$ since the range of S contains the dense subspace

$$\{T_{f,g} : f\varepsilon C_c(\Delta) , g\varepsilon C_c(\Lambda)\} \qquad (17.1).$$

17.4 THEOREM. Every K in $L^2(\Sigma,\mu\otimes\nu)$ is the kernel for an integral operator T which belongs to $\mathrm{Hom}_2(L^2(\Delta,\nu), L^2(\Lambda,\mu))$. The correspondence $K\to T$ is one-to-one and norm-preserving. It is also onto, i.e. each Hilbert-Schmidt operator from $L^2(\Delta,\nu)$ into $L^2(\Lambda,\mu)$ is an integral operator which is derived from a kernel in $L^2(\Sigma,\mu\otimes\nu)$ by this correspondence.

Proof. All that remains to be proven is that each K in $L^2(\Sigma,\mu\otimes\nu)$ is a kernel for the integral operator $T=S(K)$ in the sense of our definition (15.1). If $K \in L^2(\Sigma,\mu\otimes\nu)$, then $K \in M(\Sigma,\mu\otimes\nu)$ and hence, $K \in \mathcal{M}_{us}(\Lambda,\Delta,\mu,\nu)$ (14.6), i.e. $K \in M_w(\Lambda^*,\Delta,\mu,\nu)$ (11.10). Therefore, if and $f \in C_c(\Delta)$ and $g \in C_c(\Lambda)$, then $f \in M(\Delta,\nu)$ and $g \in M(\Lambda,\mu)$ and consequently, $<f_K:g^*\lambda>$ is $\mu\otimes\nu$-measurable. Moreover,

$$\begin{aligned}
\int|<K(x,y)f(y),g(x)>|d(x,y) &\leq \int\int\|K(x,y)\|\ \|g(x)\|\ dxdy \\
&\leq \int\|f(y)\|\int\|K(x,y)\|_s\ \|g(x)\|\ dxdy \\
&\leq \|g\|_2\int\|f(y)\|\ (\int\|K(x,y)\|_s^2\ dx)^{\frac{1}{2}}dy \\
&\leq \|f\|_2\ \|g\|_2\ (\int\int\|K(x,y)\|_s^2\ dxdy)^{\frac{1}{2}} \\
&= \|f\|_2\ \|g\|_2\ \|K\|_2 .
\end{aligned}$$

Thus, the function $<f_K:g^*\lambda>$ is $\mu\otimes\nu$-integrable.

Next let $\{K_n\}$ be a sequence in Σ_c such that $K_n\to K$ in $L^2(\Sigma,\mu\otimes\nu)$. Then $\{K_n\}$ is a kernel for the integral operator $T_n=S(K_n)$ (15.8), so that

$$<T_nf,g> = \int <f_{K_n}:g^*\lambda>\ d\mu\otimes\nu , \qquad n=1.2,\dots .$$

Moreover, $T_n\to T$ in $\mathrm{Hom}_2(L^2(\Delta,\nu)),(L^2(\Lambda,\mu))$ which implies $T_n\to T$ in $\mathrm{Hom}(L^2(\Delta,\nu),L^2(\Lambda,\mu))$. Hence,

$$<T_nf,g> \to <Tf,g> .$$

On the other hand,

$$\begin{aligned}
|\int<f_K:g^*\lambda>\ d\mu\otimes\nu - \int<f_{K_n}:g^*\lambda>\ d\mu\otimes\nu| &\leq \int|<f_{K-K_n}:g^*\lambda>|\ d\mu\otimes\nu \\
&\leq \|f\|_2\ \|g\|_2\ \|K-K_n\|_2 ,
\end{aligned}$$

as above. Thus,

$$<Tf,g> = \int<f_K:g^*\lambda>\ d\mu\otimes\nu ,$$

which completes the proof.

17.5 REMARKS. For each K in $L^2(\Sigma,\mu\otimes\nu)$, the previous inner product formula is valid for all f in $L^2(\Delta,\nu)$ and g in $L^2(\Lambda,\mu)$ (15.6), thus determining $T=S(K)$ in terms of K. In the opposite direction, given T, how do we obtain $K=S^{-1}(T)$ in terms of T? This is accomplished by a familiar Hilbert space argument (Schatten [1, p.337]). Since T is Hilbert-Schmidt, it may be written in the form

$$T = \sum_{i=1}^{\infty} a_i T_{f_i,g_i} ,$$

where $\{f_i\}$ (resp. $\{g_i\}$) is an orthonormal sequence in $L^2(\Delta,\nu)$ (resp. $L^2(\Lambda,\mu)$) and $\{a_i\}$ is a sequence of complex numbers satisfying:

$$\sum_{i=1}^{\infty} a_i^2 = \|T\|_s^2 .$$

Then:

$$S^{-1}(T_{f_i,g_i}) = K_{f_i,g_i}$$

and hence:

$$K = S^{-1}(T) = \sum_{i=1}^{\infty} a_i K_{f_i,g_i} .$$

17.6 COROLLARY. Suppose $E_x=E$ and $F_y=F$, for each x in X, y in Y. Then $\mathrm{Hom}_2(L^2(Y,\nu,F), L^2(X,\mu,E))$ is unitarily equivalent to $L^2(X\times Y,\mu\otimes\nu,\mathrm{Hom}_2(F,E))$. If T corresponds to K under this equivalence, then for each f in $L^2(Y,\nu,F)$ we have:

$$Tf(x) = \int K(x,y)f(y)dy ,$$

μ-almost everywhere on X.

Proof. The first part is well-known and also follows from 17.4 by choosing $\Lambda=\Lambda_0$ and $\Delta=\Delta_0$ as in 1.2. Since $\Sigma(\Lambda_0,\Delta_0) \sim \Sigma_0$ (14.2), we have:

$$L^2(\Delta_0,\nu) = L^2(Y,\nu,F) ,$$

$$L^2(\Lambda_0,\mu) = L^2(X,\mu,E)$$

and

$$L^2(\Sigma(\Lambda_o,\Delta_o),\mu\otimes\nu) = L^2(\Sigma_o,\mu\otimes\nu)$$
$$= L^2(X\times Y,\mu\otimes\nu,\mathrm{Hom}_2(F,E)).$$

For the second part, recall that:

$$\langle Tf,g\rangle = \int\langle f_K:g*\lambda\rangle\, d\mu\otimes\nu ,$$

for $f\in L^2(Y,\nu,F)$ and $g\in L^2(\Lambda,\mu,E)$. Also, since K is a strongly measurable mapping from $X\times Y$ into $\mathrm{Hom}(F,E)$ (14.6), it follows that the function $(x,y)\to K(x,y)f(y)$ belongs to $M(X\times Y,\mu\otimes\nu,E)$ for each f. Thus, for μ-almost all x in X, the function $y\to K(x,y)f(y)$ of Y into E is μ-measurable (Bourbaki [2, p.89]) and hence, μ-integrable (by the Hölder Inequality and the Fubini Theorem). Consequently, for μ-almost all x in X, the integral

$$\int K(x,y)f(y)dy$$

is an element of E. Moreover, by the previous equality, the resulting function

$$x \to \int K(x,y)f(y)dy$$

agrees with Tf μ-almost everywhere on X.

Next we have the following partial response to the Hilbert-Schmidt analogue of question (1) of 16.8.

17.7 PROPOSITION. Suppose T is an integral operator from $L^2(\Delta,\nu)$ into $L^2(\Lambda,\mu)$ with kernel K. If T is Hilbert-Schmidt and Λ,Δ are locally separable, then $K(x,y)$ is Hilbert-Schmidt for locally $\mu\otimes\nu$-almost all (x,y) in $X\times Y$.

Proof. If T is Hilbert-Schmidt, there exists K_1 in $L^2(\Sigma,\mu\otimes\nu)$ such that $S(K_1)=T$. In particular, K_1 is a kernel for T (17.4) and $K_1(x,y)$ is Hilbert-Schmidt for $\mu\otimes\nu$-almost all (x,y) in $X\times Y$. Hence, $K-K_1$ is a kernel for the zero operator, which implies that $K=K_1$ locally $\mu\otimes\nu$-almost everywhere (15.5).

Finally, suppose $X=Y$, $E_x=F_x$, for all x in X , $\Lambda=\Delta$ and $\mu=\nu$. Recall that $\mathrm{Hom}_2(L^2(\Lambda,\mu))$ is an H*-algebra with involution and multiplication given by the adjoint and composition operations respectively. If $K\varepsilon L^2(\Sigma,\mu\otimes\mu)$ and $T=S(K)$, then $K^*\varepsilon L^2(\Sigma^*,\mu\otimes\mu)$, where, for $\Sigma^* = \langle\cdot,\Sigma\rangle$, we have:

$$L^2(\Sigma^*,\mu\otimes\mu) \cong L^2(\Sigma,\mu\otimes\mu)^* \tag{8.13}$$

$$\cong L^2(\Sigma,\mu\otimes\mu) .$$

Moreover, considered as an element of $L^2(\Sigma, \mu\otimes\mu)$, K^* has the property that $S(K^*) = T^*$ by 15.12, since these operators have the same quadratic form. Thus, the involution in $L^2(\Sigma,\mu\otimes\mu)$ is given by $K\to K^*$. To determine the multiplication, we proceed as follows. If K_1 and K_2 are in Σ_c , then for each (x,y) in $X\times X$, the mapping

$$z \to K_1(x,z)K_2(z,y)$$

of X into $\mathrm{Hom}_2(E_y,E_x)$ is continuous with compact support. Hence, we may define K_1*K_2 by

$$K_1*K_2(x,y) = \int K_1(x,z)K_2(z,y)dz , \qquad (x,y)\varepsilon X\times X.$$

It is easily seen that K_1*K_2 is in Σ_c ,

$$\|K_1*K_2\|_2 \leq \|K_1\|_2 \|K_2\|$$

and

$$S(K_1*K_2) = S(K_1)\cdot S(K_2) .$$

Therefore, by 17.3, the mapping

$$(K_1,K_2) \to K_1*K_2$$

extends to a bilinear mapping

$$L^2(\Sigma,\mu\otimes\mu)\times L^2(\Sigma,\mu\otimes\mu) \to L^2(\Sigma,\mu\otimes\mu)$$

satisfying the previous two formulas. Thus, the kernel K_1*K_2 corresponds to the integral operator $S(K_1)\cdot S(K_2)$ in $\mathrm{Hom}_2(L^2(\Lambda,\mu))$. In the classical case, there is a familiar integral description of K_1*K_2. To extend this description to this setting, let K_1 and K_2 be elements of $L^2(\Sigma,\mu\otimes\mu)$. Omitting the details of a tedious but straightforward argument, it can be shown that for almost all (x,y) in $X\times X$, the mapping

$$X \to \mathrm{Hom}_2(E_y,E_x)$$

given by

$$z \to K_1(x,z)K_2(z,y)$$

is μ-integrable and thus defines an element

$$\int K_1(x,z)K_2(z,y)dz$$

of $\mathrm{Hom}_2(E_y,E_x)$. Furthermore, the kernel K_1*K_2 can be shown to satisfy:

$$K_1*K_2(x,y) = \int K_1(x,z)K_2(z,y)dz$$

for almost all (x,y) in $X\times X$. Hence, we have a complete description of the multiplication in $L^2(\Sigma,\mu\otimes\mu)$.

18. TRACE-CLASS OPERATORS

Our remaining concern in this chapter is the question of when an integral operator on the Hilbert space $L^2(\Lambda,\mu)$ is trace-class. In addition to finding conditions on the kernel which make the operator trace-class, we also wish to compute the trace of the operator in terms of the kernel. In this section, we present several results in response to these questions. We continue to work in the Hilbert space setting as at the end of Section 17. Throughout this section, T will denote an integral operator on $L^2(\Lambda,\mu)$ with kernel K. First we consider the question of necessity.

18.1 PROPOSITION. Suppose T is trace-class and K is the unique element of $L^2(\Sigma,\mu\otimes\mu)$ for which $S(K)=T$. If K is weakly continuous, then:

(i) $K(x,x)$ is trace-class for almost all x in X.

(ii) The function $x \to tr(K(x,x))$ is integrable.

(iii) $$tr(T) = \int tr(K(x,x)dx .$$

Proof. This is proved as in Duflo [1, Prop. 3.1.1]. Also, see Gaal[1,p.386].

18.2 THEOREM. Suppose T is trace-class and K is weakly continuous. Suppose also that either Λ is locally separable or $supp(\mu)$ is σ-compact. Then:

(i) $K(x,y)$ is Hilbert-Schmidt for almost all (x,y) in $X\times X$.

(ii) $K(x,x)$ is trace-class for almost all x in X.

(iii) The function $x \to tr(K(x,x))$ is integrable.

(iv) $$tr(T) = \int tr(K(x,x)dx .$$

Proof. First suppose that Λ is locally separable. Let K' be the unique element of $L^2(\Sigma,\mu\otimes\mu)$ such that $S(K')=T$. Since K' is also a kernel for T, $K-K'$ is a kernel for the zero operator. Thus, $K=K'$ almost everywhere on $X\times X$ (15.5) and (i)-(iv) follow from the previous proposition.

Next suppose that $supp(\mu)$ is σ-compact. Since we may assume $X = supp(\mu)$ (section 6), this part of the theorem is precisely Proposition 3.1.1 of Duflo [1].

For the question of sufficiency, we require the following preliminary results.

18.3 PROPOSITION. Suppose T is a positive operator (15.13), K is weakly continuous and

$$\int^{-} \mathrm{tr}(K(x,x))dx < \infty .$$

Then $K(x,y)$ is Hilbert-Schmidt for almost all (x,y) in $X \times X$ and

$$\int^{-} \|K(x,y)\|_s^2 \, d(x,y) \leq \int^{-} \mathrm{tr}(K(x,x))dx .$$

Proof. See Théorème 3.3.1 of Duflo [1].

18.4 THEOREM. Suppose $X = \mathrm{supp}(\mu)$, X is sigma-compact, K is weakly continuous and $(\Sigma,\mu\otimes\mu)$-measurable, T is positive (recall 14.8, 14.9, and 15.13) and

$$\int^{-} \mathrm{tr}(K(x,x))dx < \infty .$$

Then T is a trace-class operator on $L^2(\Lambda,\mu)$.

Proof. See Thèoréme 3.3.1 of Duflo [1].

18.5 REMARKS. Basically, the problem here is to find how to distinguish the trace-class operators on $L^2(\Lambda,\mu)$ from those which are Hilbert-Schmidt in terms of the corresponding kernels. More specifically, given K in $L^2(\Sigma,\mu\otimes\mu)$, what extra conditions on K guarantee that the corresponding operator T is trace-class? In view of the preceding results, one is tempted to conjecture that the extra conditions are that $K(x,x)$ be trace-class, for almost all x in X, and the resulting function $x \to \mathrm{tr}(K(x,x))$ be integrable. However, as we see next, this is false in general even for continuous K.

18.6 EXAMPLE (Communicated by R. Lipsman). Suppose G is a compact abelian group with normalized Haar measure and identity e, so that $\hat{G}$ (the unitary dual of G) is a discrete abelian group. Let k be a continuous function on G whose Fourier series is divergent at some point x_0 of G

(such exist). Since G is compact, $L^2(G) \subseteq L^1(G)$ and left convolution by k defines a bounded integral operator T on $L^2(G)$ given by

$$(Tg)(x) = \int k(xy^{-1})g(y)dy \ , \ g \in L^2(G), x \in G.$$

Thus, T has kernel K, where $K(x,y) = k(xy^{-1})$. Furthermore, $K(x,x) = k(e)$ and $\int_G tr(K(x,x))dx = k(e)$. However, T is not trace-class as the following argument verifies. The Fourier transform is a unitary isomorphism of $L^2(G)$ with $L^2(\hat{G})$. Hence, T corresponds to an operator $\hat{T}$ on $L^2(\hat{G})$ given by

$$(\hat{T}h)(\gamma) = \hat{k}(\gamma) \cdot h(\gamma), \ \gamma \in \hat{G}, \ h \in L^2(\hat{G}).$$

Consequently, T is trace-class if and only if $\hat{T}$ is. But $\hat{T}$ trace-class implies $\sum_{\hat{G}} |\hat{k}(\gamma)| < \infty$, i.e. the Fourier Series of k is absolutely convergent on G. In particular it converges at x_0, which is a contradiction.

V. APPLICATION TO INDUCED BANACH REPRESENTATIONS

As indicated in the beginning, our main purpose in these notes was to construct a foundation for the notions of kernel and integral operator in the context of continuous sums of Banach spaces. Of particular concern, were (1) the meanings and properties of the familiar types of continuity and measurability for such kernels and (2) conditions on the kernels which would insure certain compactness properties of the integral operators. Our motivation for these concerns was to be able to (1) determine the type of continuity exhibited by certain kernels associated with the integrated form of an induced Banach representation and (2) find conditions on the original representation which would guarantee that the induced representation had certain compactness properties. We are now prepared to respond to these questions. We shall see that the kernels of interest are (essentially) continuous in the strongest sense discussed in Chapter IV, namely norm-continuous. We will also give some conditions on an "inducible" representation which imply that the induced representation is compact, Hilbert-Schmidt or trace-class.

19. INDUCED BANACH REPRESENTATIONS

We begin by defining what we mean by an induced Banach representation. The discussion and results of this section are a summary of section 3 of Fontenot and Schochetman [1].

Suppose G is a locally compact group and H is a closed subgroup of G. If Δ_G and Δ_H denote the (right) modular functions of G and H respectively, let δ denote the quotient function $\Delta_H/(\Delta_G|H)$. Also, X will denote the space G/H of right cosets and π the projection of G onto X.

Corresponding to a continuous "rho-function" $\rho:G \to (o,\infty)$, there is a positive Radon measure μ on X (a quasi-invariant measure) whose support is all of X. As before, we will write $d(\pi(x))$ for $d\mu(\pi(x))$.

Now let L be a strongly continuous representation of H on the Banach space E. Then L is Banach inducible up to G if there exists an inducing pair for L up to G , i.e. a pair (p,q) in $(1,\infty)\times(o,\infty)$ satisfying:

$$\delta(t)^{1/q} \|L(t)\| \leq b\delta(t)^{1/p} , \qquad t\varepsilon H ,$$

where $b\geq 1$. For our purposes here, we restrict our attention to the case $1<p,q<\infty$. Given such a pair, we can construct an isometric representation of G on some Banach space as follows. Let $C_q(G,L)$ denote the space of continuous functions $f:G\to E$ having compact support modulo H (i.e. $\pi(\text{supp}(f))$ is compact) and satisfying:

$$f(tx) = \delta(t)^{1/q} L(t)f(x) , \qquad t\varepsilon H, x\varepsilon G.$$

For each such f , define:

$$N_p(f,x) = \rho(x)^{-1/p} \sup_{t\varepsilon H} \{ \delta(t)^{-1/p}\|f(tx)\| \} , \qquad x\varepsilon G ,$$

and

$$N_p(f) = (\int_X N_p(f,x)^p \, d(\pi(x)))^{1/p}.$$

(Note that the function $N_p(f,\cdot)$ is constant on cosets and lower semi-continuous.) Then $C_q(G,L)$ is a normed linear space with norm given by N_p. Its completion $F^p_q(G,L)$ is thus a Banach space.

The right translation mapping $f\to f_x$ by an element x of G is an isometry of $C_q(G,L)$ and hence extends to an isometry of $F^p_q(G,L)$ which we denote by $U(x)$. (Of course, $U(x)$ depends on p,q and L.) The resulting mapping $U:x \to U(x)$ is then a strongly continuous (isometric) representation of G on $F^p_q(G,L)$.

Before going on to study the integrated form of U, we observe the following. If we let $M = \delta^{1/q-1/p}L$, then M is also a representation of H on E --but M is bounded by b. Therefore, we may renorm E by defining:

$$\|v\|_M = \sup_{t \in H} \|M(t)v\| ,$$

so that

$$\|v\| \leq \|v\|_M \leq b\|v\| , \qquad v \in E.$$

If E_M denotes E with norm replaced by $\|\cdot\|_M$, then E_M is a Banach space which is equivalent to E. Furthermore,

$$N_p(f,x) = \rho(x)^{-1/p}\|f(x)\|_M , \qquad x \in G ,$$

and

$$N_p(f) = (\int_X \rho(x)^{-1}\|f(x)\|_M^p \, d(\pi(x)))^{1/p} .$$

20. THE INTEGRATED FORM

In this section we show that the operators of the integrated form of U can be written as strong integral operators on an L^p-continuous sum of Banach spaces which resemble E. The most significant fact here is that the corresponding kernel fields are norm-continuous. The results stated here can be found in Chapter II of Schochetman [1].

The integrated form of U is the representation (also denoted by U) on $F_q^p(G,L)$ of the group algebra $L^1(G)$ obtained by defining:

$$U(\phi) = \int_G \phi(x)U(x)dx , \qquad \phi\varepsilon C_c(G) ,$$

where dx denotes a right Haar measure on G. (For our purposes, it suffices to work with $C_c(G)$ instead of $L^1(G)$.) Thus, for f in $C_q(G,L)$, we have:

$$(U(\phi)f)(x) = \int_G \phi(y)U(y)f(x)dy ,$$

$$= \int_G \phi(y)f(xy)dy , \qquad x\varepsilon G ,$$

which may be further decomposed as follows:

$$(U(\phi)f)(x) = \int_X I_\phi(x,y)f(y)d(\pi(y)) ,$$

(again the integrand is constant on cosets) where I_ϕ is a continuous mapping of $G\times G$ into Hom(E) given by:

$$I_\phi(x,y) = \Delta_G(x^{-1})\rho(y)^{-1}\int_H \delta(t)^{1/q-1}\phi(x^{-1}ty)L(t)dt , \qquad x,y\varepsilon G ,$$

where dt is a right Haar measure on H. Therefore, the operators $U(\phi)$ resemble integral operators and their corresponding "kernels" I_ϕ are uniformly continuous. These are the facts which suggest that the integrated form of U should consist of integral operators with uniformly continuous kernels. To formalize this statement, we must redevelop $F_q^p(G,L)$ and the

$U(\phi)$ (ϕ in $C_c(G)$) along the lines of Chapters I-IV. This is our goal in the remainder of this section.

20.1 PROPOSITION. For each ξ in X and y in G such that $\pi(y)=\xi$, there exists a Banach space E_ξ equivalent to E whose norm $\|\cdot\|_\xi$ satisfies:

$$\|\cdot\|_M \leq \|\cdot\|_\xi \leq \rho(y)^{1/q-1/p}\|\cdot\|_M ,$$

if $\rho(y)^{1/q-1/p} \geq 1$, and

$$\rho(y)^{1/q-1/p}\|\cdot\|_M \leq \|\cdot\|_\xi \leq \|\cdot\|_M ,$$

if $\rho(y)^{1/q-1/p} \leq 1$. Furthermore, each E_ξ is a Hilbert space if and only if E_M is a Hilbert space.

Hence, we have a field $\{E_\xi\}$ of (equivalent) Banach spaces over X. These spaces need not be isometrically equivalent in general (they will be if $p=q$). Next we need a continuity structure in ΠE_ξ relative to which $F_q^p(G,L)$ becomes a Lebesque space as in §6.

20.2 PROPOSITION. There exists a continuity structure Λ_q in ΠE_ξ satisfying the following:

(i) Λ_q is countably dense if and only if it is separable; in fact, $C(\Lambda_q) = \Lambda_q$.

(ii) If E is separable and G is second countable, then Λ_q is countably dense.

20.3 PROPOSITION. There exists an isometry S of $C_q(G,L)$ onto $C_c(\Lambda_q)$. Therefore, $F_q^p(G,L)$ is isometrically isomorphic to $L^p(\Lambda_q,\mu)$. Moreover, $F_q^p(G,L)$ is a Hilbert space if and only if $p=2$ and E_M is a Hilbert space.

The next step is to transfer U from $F_q^p(G,L)$ to $L^p(\Lambda_q,\mu)$. To do this we define:

$$W(x) = SU(x)S^{-1} \qquad x\varepsilon G ,$$

thereby obtaining a representation W of G on $L^p(\Lambda_q,\mu)$ which is isometrically equivalent to U. The integrated form of W is precisely the collection of operators we are after.

20.4 THEOREM. For each ϕ in $C_c(G)$, there exists a kernel K_ϕ in $\Pi\mathrm{Hom}(E_\eta,E_\xi)$ such that: For each f in $C_c(\Lambda_q)$, the function $\eta\to K_\phi(\xi,\eta)f(\eta)$ of X into E_ξ is μ-integrable, and

$$(W(\phi)f)(\xi) = \int_X K_\phi(\xi,\eta)f(\eta)d\eta , \qquad \xi\varepsilon X.$$

Therefore, $W(\phi)$ is strongly an integral operator on $L^p(\Lambda_q,\mu)$ with kernel K_ϕ.

Finally, we have the following result regarding the continuity of K_ϕ, $\phi\varepsilon C_c(G)$.

20.5 THEOREM. There exists a pre-continuity structure Ω in $\Pi\mathrm{Hom}(E_\eta,E_\xi)$ with respect to which K_ϕ is continuous, i.e. $K_\phi\varepsilon C(\Omega)$. The structure Ω is a strong (Λ,Λ)-pre-continuity structure, i.e. $\Omega \subseteq \mathcal{C}_s(\Lambda,\Lambda)$ (recall 12.20).

20.6 REMARK. Technically, the previous theorem does not guarantee that the K_ϕ are norm-continuous (or uniformly continuous) in the sense of section 12. This is a consequence of $\Omega \subseteq \mathcal{C}_s(\Lambda,\Lambda)$ as opposed to $\Omega \subseteq \mathcal{C}_u(\Lambda,\Lambda)$. However, it is "almost" the case that $\Omega \subseteq \mathcal{C}_u(\Lambda,\Lambda)$, i.e. Ω "is" a uniform (Λ,Λ)-pre-continuity structure as much as it can be expected to be under the existing circumstances. These details and their explanations can be found in Schochetman [1].

21. COMPACTNESS FOR INDUCED REPRESENTATIONS

Our last consideration in these notes is the application of Chapter IV to the integrated form of U. The following are (essentially) translations of the main results of this chapter to the setting of the "integral operators" $U(\phi)$, $\phi\epsilon C_c(G)$. For the details, see Schochetman [1, Chap. IV].

For the purpose of the following result, it can be shown that:

$$\|K_\phi(\xi,\eta)\| \leq b\rho(x)^{-1/q}\ \rho(y)^{1/q}\ \|I_\phi(x,y)\| \ ,$$

for ξ,η in X and x,y in G such that $x\epsilon\xi$, $y\epsilon\eta$.

21.1 THEOREM. Suppose the representation L is such that $L(\beta)$ is a compact operator on E for each β in $C_c(H)$. Let $\phi\epsilon C_c(G)$. If

$$\left(\int_X\left(\int_X \rho(x)^{-p^*/q}\rho(y)^{p^*/q}\ \|I_\phi(x,y)\|^{p^*}d\eta\right)^{p/p^*}d\xi\right)^{1/p}<\infty$$

(the integrand is constant on cosets), then $U(\phi)$ is a compact operator on $F^p_q(G,L)$.

Proof. Apply 16.4 with q=p. Note that the compactness condition of L insures that Ω can be chosen to be of compact type and the integral condition on I_ϕ insures that $K_\phi\sigma$ belongs to $L^{p^*,p}(\Omega\sigma,\mu,\mu)$.

21.2 COROLLARY. If X is compact and L is as in the theorem, then $U(\phi)$ is a compact operator for each ϕ in $C_c(G)$.

Next we turn to the Hilbert space context. Accordingly, assume p=2 and E_M , i.e. each E_ξ, $\xi\epsilon X$, is a Hilbert space, so that $L^2(\Lambda_q,\mu)$ is a Hilbert space. If each $I_\phi(x,y)$ is a Hilbert-Schmidt operator on E_M , then so is each $K_\phi(\xi,\eta)$ and

$$\|K_\phi(\xi,\eta)\|_s = \rho(x)^{-1/q}\rho(y)^{1/q}\ \|I_\phi(x,y)\|_s \ ,$$

for ξ,η in X and x,y in G such that $x\varepsilon\xi, y\varepsilon\eta$.

21.3 THEOREM. Suppose G is second countable, E is separable, and $L(\beta)$ is a Hilbert-Schmidt operator on E_M for each β in $C_c(H)$. Let $\phi\varepsilon C_c(G)$. If

$$\left(\int_{X\times X}\rho(x)^{-2/q}\rho(y)^{2/q}\,\|I_\phi(x,y)\|_s^2\,d(\xi,\eta)\right)^{1/2} < \infty$$

(the integrand is constant on cosets), then $U(\phi)$ is a Hilbert-Schmidt operator on $F_q^2(G,L)$ and conversely.

Proof. Apply 17.4. Note that the Hilbert-Schmidt property of L insures that K_ϕ is Hilbert-Schmidt-valued while the separability of E and second countability of G guarantee that K_ϕ is $(\Sigma(\Lambda_q),\mu\otimes\nu)$-measurable (14.8).

21.4 COROLLARY. Suppose X is compact, L is as in the theorem, and $L:C_c(H) \to \mathrm{Hom}_2(E_M)$ is continuous (relative to the inductive limit topology on the domain space). Then $U(\phi)$ is a Hilbert-Schmidt operator for each ϕ in $C_c(G)$.

Finally, for the following result, observe that each $K_\phi(\xi,\xi)$ is trace-class if each $I_\phi(x,x)$ is and

$$\mathrm{tr}(K_\phi(\xi,\xi)) = \mathrm{tr}(I_\phi(x,x)) ,$$

for $\xi\varepsilon X$ and $x\varepsilon\xi$.

21.5 THEOREM. Suppose G is second countable, and E is separable. Let $\phi\varepsilon C_c(G)$ be such that $U(\phi)$ is a positive operator on $F_q^2(G,L)$. (For example, such is the case if ϕ is a positive element of the group C*-algebra of G.) If

$$\int_X \mathrm{tr}(I_\phi(x,x))\,d\xi < \infty$$

(the integrand is constant on cosets), then $U(\phi)$ is a trace-class operator on $F_q^2(G,L)$ and

$$tr(U(\phi)) = \int_X tr(I_\phi(x,x))d\xi.$$

Proof. Apply 18.4 and 18.2. Note that K_Φ is automatically weakly continuous.

21.6 COROLLARY. Suppose X is compact and the mapping $L:C_c(H) \to Hom_2(E_M)$ is such that $\beta \to tr(L(\beta))$ is a continuous function on $C_c(H)$. Then $U(\phi)$ is a trace-class operator for each ϕ in $C_c(G)$ such that $U(\phi)$ is a positive operator (with trace as in the theorem).

REFERENCES

A. BENEDEK and R. PANZONE
1. The spaces L^p with mixed-norm, Duke J. Math., 28 (1961), 301-324.

N. BOURBAKI
1. Intégration, Chap. 1, 2, 3, et 4, deuxième éd., Hermann, Paris, 1965.
2. Intégration, Chap. 5, deuxième éd., Hermann, Paris, 1967.

N. DINCULEANU
1. Sur la représentation intégrale des certaines opérations linéaires II, Comp. Math. 14 (1959), 1-22.

J. DIXMIER
1. Les C*-algébres et leurs représentations, Gauthiers-Villars, Paris, 1964.

M. DUFLO
1. Généralités sur les représentations induites, Chap. V of Représentations des groupes de Lie résolubles, P. Bernat et al, Dunod, Paris, 1972.

N. DUNFORD and J. T. Schwartz
1. Linear Operators, Part I, Interscience, New York, 1964.

J. M. G. FELL
1. The structure of algebras of operator fields, Acta Math. 106 (1961), 233-280.
2. Large compact subgroups of locally compact groups, unpublished notes.

R. A. FONTENOT and I. SCHOCHETMAN
1. Induced representations of groups on Banach spaces, Rocky Mt. J. Math, 7 (1977), 53-82.

R. GODEMENT
1. Sur la théorie des représentations unitaires, Ann. Math. 53 (1951), 68-124.

E. HEWITT and K. STROMBERG
1. Real and abstract analysis, Springer-Verlag, New York, 1965.

R. SCHATTEN
1. Norm ideals of completely continuous operators, Springer-Verlag, 1970.

I. SCHOCHETMAN
1. Integral operators in the theory of Induced Banach representations, submitted for publication.
2. Topology and the duals of certain locally compact groups, Trans. A.M.S. 150 (1970), 477-489.

Department of Mathematical Sciences
Oakland University
Rochester, Michigan 48063